Student Solutions Manual

Basic Mathematics for College Students

FOURTH EDITION

Alan S. Tussy
Citrus College

R. David Gustafson
Rock Valley College

Diane R. Koenig
Rock Valley College

Prepared by

Nathan G. Wilson
St. Louis Community College at Meramec

BROOKS/COLE
CENGAGE Learning

Australia • Brazil • Japan • Korea • Mexico • Singapore • Spain • United Kingdom • United States

© 2011 Brooks/Cole, Cengage Learning

ALL RIGHTS RESERVED. No part of this work covered by the copyright herein may be reproduced, transmitted, stored, or used in any form or by any means graphic, electronic, or mechanical, including but not limited to photocopying, recording, scanning, digitizing, taping, Web distribution, information networks, or information storage and retrieval systems, except as permitted under Section 107 or 108 of the 1976 United States Copyright Act, without the prior written permission of the publisher.

For product information and technology assistance, contact us at **Cengage Learning Customer & Sales Support, 1-800-354-9706**

For permission to use material from this text or product, submit all requests online at **www.cengage.com/permissions** Further permissions questions can be emailed to **permissionrequest@cengage.com**

ISBN-13: 978-0-538-73408-0
ISBN-10: 0-538-73408-6

Brooks/Cole
20 Davis Drive
Belmont, CA 94002-3098
USA

Cengage Learning is a leading provider of customized learning solutions with office locations around the globe, including Singapore, the United Kingdom, Australia, Mexico, Brazil, and Japan. Locate your local office at: **www.cengage.com/global**

Cengage Learning products are represented in Canada by Nelson Education, Ltd.

To learn more about Brooks/Cole, visit
www.cengage.com/brookscole

Purchase any of our products at your local college store or at our preferred online store
www.cengagebrain.com

Printed in the United States of America
1 2 3 4 5   14 13 12 11 10

ED189

Table of Contents

Chapter 1: Whole Numbers
1.1 An Introduction to the Whole Numbers — 1
1.2 Adding Whole Numbers — 3
1.3 Subtracting Whole Numbers — 5
1.4 Multiplying Whole Numbers — 7
1.5 Dividing Whole Numbers — 9
1.6 Problem Solving — 13
1.7 Prime Factors and Exponents — 14
1.8 The Least Common Multiple and the Greatest Common Factor — 16
1.9 Order of Operations — 18
CHAPTER REVIEW — **22**
CHAPTER TEST — **26**

Chapter 2: The Integers
2.1 An Introduction to the Integers — 28
2.2 Adding Integers — 30
2.3 Subtracting Integers — 32
2.4 Multiplying Integers — 35
2.5 Dividing Integers — 37
2.6 Order of Operations and Estimation — 40
CHAPTER REVIEW — **44**
CHAPTER TEST — **47**
CUMULATIVE REVIEW — **48**

Chapter 3: Fractions and Mixed Numbers
3.1 An Introduction to Fractions — 51
3.2 Multiplying Fractions — 53
3.3 Dividing Fractions — 56
3.4 Adding and Subtracting Fractions — 59
3.5 Multiplying and Dividing Mixed Numbers — 63
3.6 Adding and Subtracting Mixed Numbers — 66
3.7 Order of Operations and Complex Fractions — 70
CHAPTER REVIEW — **75**
CHAPTER TEST — **79**
CUMULATIVE REVIEW — **80**

Chapter 4: Decimals
4.1 An Introduction to Decimals — 82
4.2 Adding and Subtracting Decimals — 84
4.3 Multiplying Decimals — 86
4.4 Dividing Decimals — 90
4.5 Fractions and Decimals — 95
4.6 Square Roots — 100
CHAPTER REVIEW — **102**
CHAPTER TEST — **107**
CUMULATIVE REVIEW — **108**

Chapter 5: Ratio, Proportion, and Measurement
5.1 Ratios — 111
5.2 Proportions — 113
5.3 American Units of Measurement — 118
5.4 Metric Units of Measurement — 121
5.5 Converting between American and Metric Units — 123

CHAPTER REVIEW	**126**
CHAPTER TEST	**129**
CUMULATIVE REVIEW	**130**

Chapter 6: Percent
6.1 Percents, Decimals, and Fractions	132
6.2 Solving Percent Problems Using Percent Equations and Proportions	134
6.3 Applications of Percent	138
6.4 Estimation with Percent	141
6.5 Interest	143
CHAPTER REVIEW	**146**
CHAPTER TEST	**148**
CUMULATIVE REVIEW	**149**

Chapter 7: Graphs and Statistics
7.1 Reading Graphs and Tables	151
7.2 Mean, Median, and Mode	153
CHAPTER REVIEW	**155**
CHAPTER TEST	**156**
CUMULATIVE REVIEW	**156**

Chapter 8: An Introduction to Algebra
8.1 The Language of Algebra	159
8.2 Simplifying Algebraic Expressions	160
8.3 Solving Equations Using Properties of Equality	163
8.4 More about Solving Equations	168
8.5 Using Equations to Solve Application Problems	172
8.6 Multiplication Rules for Exponents	176
CHAPTER REVIEW	**177**
CHAPTER TEST	**181**
CUMULATIVE REVIEW	**183**

Chapter 9: An Introduction to Geometry
9.1 Basic Geometric Figures, Angles	187
9.2 Parallel and Perpendicular Lines	189
9.3 Triangles	192
9.4 The Pythagorean Theorem	195
9.5 Congruent Triangles and Similar Triangles	197
9.6 Quadrilaterals and Other Polygons	200
9.7 Perimeters and Areas of Polygons	202
9.8 Circles	206
9.9 Volume	207
CHAPTER REVIEW	**210**
CHAPTER TEST	**213**
CUMULATIVE REVIEW	**214**

Appendix II Polynomials	**219**
Appendix III Inductive and Deductive Reasoning	**224**

CHAPTER 1 Whole Numbers

Section 1.1: An Introduction to the Whole Numbers

VOCABULARY

1. The numbers 0, 1, 2, 3, 4, 5, 6, 7, 8, and 9 are the digits.

3. When we write five thousand eighty-nine as 5,089, we are writing the number in standard form.

5. When 297 is written as 200 + 90 + 7, we are writing 297 in expanded form.

7. The symbols < and > are inequality symbols.

CONCEPTS

9.

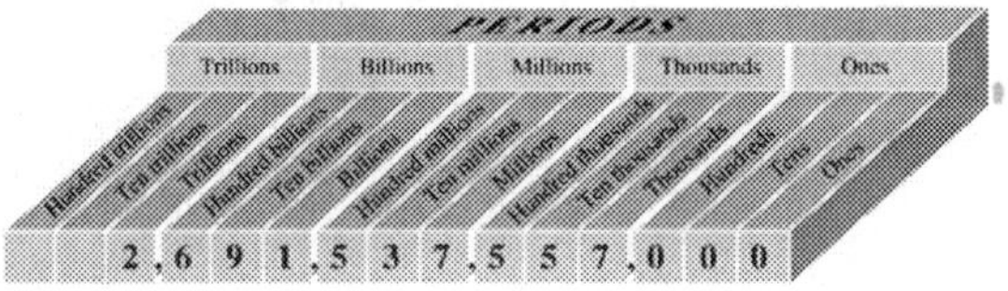
PERIODS
Trillions
Billions
Millions
Thousands
Ones
2,691,537,557,000

11. a. forty
b. ninety
c. sixty-eight
d. fifteen

13.

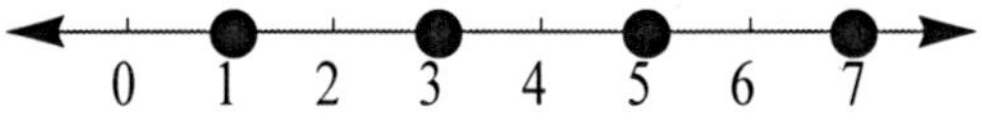
0 1 2 3 4 5 6 7

15.

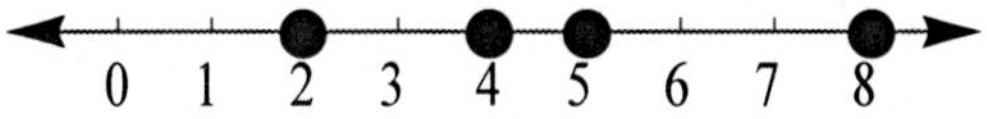
0 1 2 3 4 5 6 7 8

17.

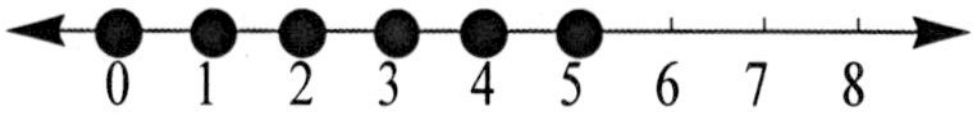
0 1 2 3 4 5 6 7 8

19.

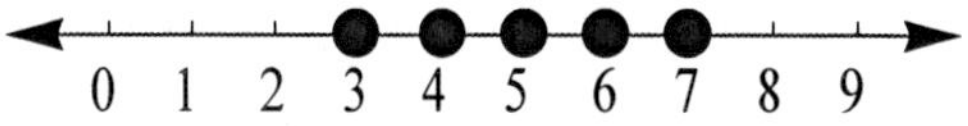
0 1 2 3 4 5 6 7 8 9

NOTATION

21. The symbols { }, called braces are used when writing a set.

GUIDED PRACTICE

23. a. 3 tens
b. 7
c. 6 hundreds
d. 5

25. a. 1 hundred million
b. 7
c. 9 tens
d. 4

27. 93 = ninety-three

29. 732 = seven hundred thirty-two

31. 154,302 = one hundred fifty-four thousand, three hundred two

33. 14,432,500 = fourteen million, four hundred thirty-two thousand, five hundred

35. 970,031,500,104 = nine hundred seventy billion, thirty-one million, five hundred thousand, one hundred four

37. 82,000,415 = eighty-two million, four hundred fifteen

39. 3,737

41. 930

43. 7,021

45. 26,000,432

© 2011 Cengage Learning. All Rights Reserved. May not be scanned, copied or duplicated, or posted to a publicly accessible website, in whole or in part.

47. $245 = 200 + 40 + 5$

49. $3,609 = 3,000 + 600 + 9$

51. $72,533 = 70,000 + 2,000 + 500 + 30 + 3$

53. $104,401 = 100,000 + 4,000 + 400 + 1$

55. $8,403,613 = 8,000,000 + 400,000 + 3,000 + 600 + 10 + 3$

57. $26,000,156 = 20,000,000 + 6,000,000 + 100 + 50 + 6$

59. a. $11 > 8$ b. $29 < 54$

61. a. $12,321 > 12,209$ b. $23,223 < 23,231$

63. 98,150, since $4 < 5$

65. 512,970, since $7 \geq 5$

67. 8,400, since $5 \geq 5$

69. 32,400, since $3 < 5$

71. 66,000, (since $8 \geq 5$, 981 rounds to 1,000)

73. 2,581,000, (since $5 \geq 5$, 952 rounds to 1,000)

75. 53,000 ; 50,000

77. 77,000 ; 80,000

79. 816,000 ; 820,000

81. 297,000 ; 300,000

83. a. 79,590 b. 79,600
    c. 80,000 d. 80,000

85. a. $419,160 b. $419,200
    c. $419,000 d. $420,000

87. 40,025

89. 202,036

91. 27,598

93. 10,700,506

APPLICATIONS

95. Aisha is the closest to $4,745 without being over.

97. a. The 1970s, with 7 successful missions
    b. The 1960s, with 9 unsuccessful missions
    c. The 1960s, with 12 missions
    d. The 1980s

99.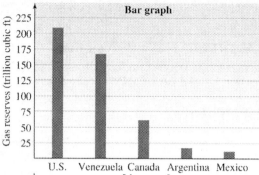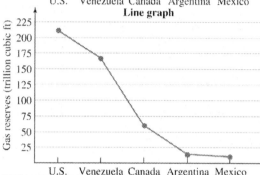

101.

103. 1,865,593 ; 482,880; 1,503; 269; 43,449

105. a. hundred thousands
     b. 980,000,000; 9 hundred millions + 8 ten millions
     c. 1,000,000,000; one billion

WRITING

107. To round 687 to the nearest ten, look to the right of the tens place. Since this digit is 7, increase the 8 in the tens place to 9 and make the ones place a zero. So, to the nearest ten, 687 is approximately 690.

109. Because **1,**000 (3 zeros) is a thousand 1s, so **1,000,**000 is a thousand thousands.

111. 2, 10, 0, 1,000, 80
     12, 3, 100, 2, 0

Section 1.2: Adding Whole Numbers

VOCABULARY

1. $\underset{\text{addend}}{10} + \underset{\text{addend}}{15} = \underset{\text{sum}}{25}$

3. The <u>commutative</u> property of addition states that the order in which whole numbers are added does not change their sum.

5. To see whether the result of an addition is reasonable, we can round the addends and <u>estimate</u> the sum.

7. The figure on the left is an example of a <u>rectangle</u>. The figure on the right is an example of a <u>square</u>.

9. When all the sides of a rectangle are the same length, we call the rectangle a <u>square</u>.

CONCEPTS

11. a. commutative property of addition
    b. associative property of addition
    c. associative property of addition
    d. commutative property of addition

13. Any number added to <u>0</u> stays the same.

NOTATION

15. The addition symbol + is read as "<u>plus</u>".

17. 33 plus 12 equals 45

19. $(36+11)+5 = 47+5$
    $ = 52$

GUIDED PRACTICE

21. $\begin{array}{r} 25 \\ +13 \\ \hline 38 \end{array}$

23. $\begin{array}{r} 406 \\ +283 \\ \hline 689 \end{array}$

25. $\begin{array}{r} 21 \\ 31 \\ +24 \\ \hline 76 \end{array}$

27. $\begin{array}{r} 603 \\ 152 \\ +121 \\ \hline 876 \end{array}$

29. $\begin{array}{r} 1 \\ 19 \\ +16 \\ \hline 35 \end{array}$

31. $\begin{array}{r} 1 \\ 45 \\ +47 \\ \hline 92 \end{array}$

33. $\begin{array}{r} 1 \\ 52 \\ +18 \\ \hline 70 \end{array}$

35. $\begin{array}{r} 1 \\ 28 \\ +47 \\ \hline 75 \end{array}$

37. $\begin{array}{r} 1 \\ 156 \\ +305 \\ \hline 461 \end{array}$

39. $\begin{array}{r} 111 \\ 4,301 \\ 789 \\ +3,847 \\ \hline 8,937 \end{array}$

41. $\begin{array}{r} 222 \\ 9,758 \\ 586 \\ +7,799 \\ \hline 18,143 \end{array}$

43. $\begin{array}{r} 23 \\ 346 \\ 217 \\ 568 \\ +679 \\ \hline 1,810 \end{array}$

45. $(9+3)+7 = 9+(3+7) = 9+10 = 19$

47. $(13+8)+12 = 13+(8+12) = 13+20 = 33$

49. $94+(6+37)$
$= (94+6)+37$
$= 100+37 = 137$

51. $125+(75+41) =$
$(125+75)+41 =$
$200+41 = 241$

53. $4+8+16+1+1$
$= (4+16)+(8+1+1)$
$= 20+10$
$= 30$

55. $23+5+7+15+10$
$= (23+7)+(5+15)+10$
$= 30+20+10$
$= 60$

57.
$$\begin{array}{r} \overset{1\,1}{624} \\ 905 \\ +\ \ 86 \\ \hline 1{,}615 \end{array}$$

59.
$$\begin{array}{r} \overset{2\,1}{457} \\ 97 \\ +\ \ 653 \\ \hline 1{,}207 \end{array}$$

61.
$$\begin{array}{r} \overset{1}{}700 \\ 800 \\ 10{,}000 \\ 20{,}000 \\ +\ \ 6{,}000 \\ \hline 37{,}500 \end{array}$$

63.
$$\begin{array}{r} 600{,}000 \\ 20{,}000 \\ 300{,}000 \\ +\ \ 100{,}000 \\ \hline 1{,}020{,}000 \end{array}$$

65. $32+12+32+12 = 88\ ft$

67. $17+17+17+17 = 68\ in$

69. $94+94+94+94 = 376\ mi$

71. $87+6+87+6 = 186\ cm$

TRY IT YOURSELF

73.
$$\begin{array}{r} 8{,}539 \\ +\ \ 7{,}368 \\ \hline 15{,}907 \end{array}$$

75.
$$\begin{array}{r} 51{,}246 \\ 578 \\ 37 \\ +\ \ 4{,}599 \\ \hline 56{,}460 \end{array}$$

77. $(45+16)+4 = 45+20 = 65$

79.
$$\begin{array}{r} 632 \\ +347 \\ \hline 979 \end{array}$$

81. $16{,}427+13{,}573 = 30{,}000$

83.
$$\begin{array}{r} 76 \\ +45 \\ \hline 121 \end{array}$$

85.
$$\begin{array}{r} 3{,}156 \\ 1{,}578 \\ +\ \ 6{,}578 \\ \hline 11{,}312 \end{array}$$

87. $12+1+8+4+9+16$
$=(12+8)+(1+9)+(4+16)$
$=20+10+20$
$=50$

APPLICATIONS

89. 24 + 35 + 16 + 16 = 91 ft.

91. 540 + 230 + 160 + 210 = 1,140 calories

93. 61,715,000 + 18,072,000 = 79,787,000 visitors

95.

Number of safe bridges	Number of bridges that need repair	Number of outdated bridges that should be replaced	Total number of bridges
445,396	72,033	80,447	597,876

97. 2,293+4,794+3,246+1,733+2,818+361+1,085+12,470 = $28,800

99. $1,036 + 1,987 + 2,202 + 1,420 = 6,645$
$6,645,000,000

101. 64 + 34 + 64 + 34 = 196 inches of fringe.

103. 24 + 24 + 24 + 24 = 96ft once around
96 + 96 + 96 + 96 = 384ft four times around

WRITING

105. Because taking 3 of something and then 4 more is the same as taking 4 of something and then 3 more – the result is the same.

107. Benefit : faster ; Tradeoff : less accurate

REVIEW

109. a. 3,000 + 100 + 20 + 5
b. 60,000 + 30 + 7

Section 1.3: Subtracting Whole Numbers

VOCABULARY

1. $\underset{\text{minuend}}{25} - \underset{\text{subtrahend}}{10} = \underset{\text{difference}}{15}$

3. The words *fall, lose, reduce,* and *decrease* often indicate the operation of subtraction.

5. To see whether the result of a subtraction is reasonable, we can round the minuend and subtrahend and estimate the difference.

CONCEPTS

7. The subtraction 7 – 3 = 4 is related to the addition statement 4 + 3 = 7.

9. To *evaluate* (find the value of) an expression that contains both addition and subtraction, we perform the operations as the occur from left to right.

NOTATION

11. The subtraction symbol – is read as "minus".

13. 83 – 30 is correct.

GUIDED PRACTICE

15. 37
−14
23

17. 89
−28
61

19. 596
−372
224

21. 674
−371
303

23. 7,989
−347
7,642

25. 2,967
−405
2,562

27.
$$\begin{array}{r}\overset{4\ 13}{\cancel{5}\,\cancel{3}}\\-1\ 7\\\hline 3\ 6\end{array}$$

29.
$$\begin{array}{r}\overset{8\ 16}{\cancel{9}\,\cancel{6}}\\-4\ 8\\\hline 4\ 8\end{array}$$

31.
$$\begin{array}{r}8,\overset{6\ \overset{13}{\cancel{7}}\ \overset{16}{\cancel{4}}\ \cancel{6}}{}\\-\ \ 2\ 8\ 9\\\hline 8,4\ 5\ 7\end{array}$$

33.
$$\begin{array}{r}6,\overset{8\ \overset{15}{\cancel{9}}\ \overset{11}{\cancel{6}}\ \cancel{1}}{}\\-\ \ 4\ 7\ 8\\\hline 6,4\ 8\ 3\end{array}$$

35.
$$\begin{array}{r}5\overset{3\ \overset{14}{\cancel{4}}\ \overset{9}{\cancel{5}}\ \overset{16}{\cancel{0}}\ \cancel{6}}{}\\-\ \ 2,8\ 2\ 9\\\hline 5\ 1,6\ 7\ 7\end{array}$$

37.
$$\begin{array}{r}4\overset{7\ \overset{13}{\cancel{8}}\ \overset{9}{\cancel{4}}\ \overset{10}{\cancel{0}}\ \overset{12}{\cancel{2}}}{}\\-\ \ 3,9\ 5\ 8\\\hline 4\ 4,4\ 4\ 4\end{array}$$

39. $123+175=298$ ☺

41. $1,364+3,275=4,629$ ☺

43.
$$\begin{array}{r}70,000\\-\ \ 4,000\\\hline 66,000\end{array}$$

45.
$$\begin{array}{r}80,000\\-30,000\\\hline 50,000\end{array}$$

47. $35-12+6=23+6=29$

49. $56-31+12=25+12=37$

51. $574+47-13=621-13=608$

53. $966+143-61=1,109-61=1,048$

TRY IT YOURSELF

55. $416-357=59$

57.
$$\begin{array}{r}3,430\\-\ \ 529\\\hline 2,901\end{array}$$

59. $301-199=102$

61.
$$\begin{array}{r}367\\-347\\\hline 20\end{array}$$

63. $633-598+30=35+30=65$

65. $420-390=30$

67. $20,007-78=19,929$

69. $852-695+40=157+40=197$

71.
$$\begin{array}{r}17,246\\-\ \ 6,789\\\hline 10,457\end{array}$$

73.
$$\begin{array}{r}15,700\\-15,397\\\hline 303\end{array}$$

75.
$$\begin{array}{r}50,009\\-\ \ 1,249\\\hline 48,760\end{array}$$

77. $120+30-40=150-40=110$

79.
$$\begin{array}{r}167,305\\-\ \ 23,746\\\hline 143,559\end{array}$$

81. $29{,}307 - 10{,}008 = 19{,}299$

APPLICATIONS

83. $1{,}689 - 269 = 1{,}420$ lbs.

85. $19{,}396 - 16{,}735 = 2{,}661$ bulldogs

87. $71{,}649 - 70{,}154 = 1{,}495$ miles

89. $63 - 8 = \$55$ fare

91. $8{,}305 - 8{,}272 = 33$ points gained

93. $1{,}947 - 183 = 1{,}764°F$

95. $26 - 9 = 17$ area codes

97. $1{,}370 - 197 + 340 = 1{,}173 + 340 = \$1{,}513$

99. a. $\$39{,}565$

    b. $40{,}887 - 39{,}565 = \$1{,}322$

WRITING

101. It is because taking 2 things from 3 things is not equivalent to taking 3 things from 2 things.

103. By adding the difference to the subtrahend, you should get the minuend.

REVIEW

105. a. $5{,}370{,}650$

     b. $5{,}370{,}000$

     c. $5{,}400{,}000$

107. $13 + 13 + 13 + 13 = 52$ in.

109.
$$
\begin{array}{r}
\overset{1\ 1\ 1}{345} \\
4{,}672 \\
+\ \ 513 \\
\hline
5{,}530
\end{array}
$$

Section 1.4: Multiplying Whole Numbers

VOCABULARY

1. $\underset{\text{factor}}{5} \cdot \underset{\text{factor}}{10} = \underset{\text{product}}{50}$

3. The <u>commutative</u> property of multiplication states that the order in which whole numbers are multiplied does not change their product. The <u>associative</u> property of multiplication states that the way in which whole numbers are grouped does not change their product.

5. If a square measures 1 inch on a side, its area is 1 <u>square</u> inch.

CONCEPTS

7. a. $4 \cdot 8$

   b. $15 + 15 + 15 + 15 + 15 + 15 + 15$

9. a. 3

   b. 5

11. a. area

    b. perimeter

    c. area

    d. perimeter

NOTATION

13. $\times\ ,\ \cdot\ ,\ (\ )$

15. $A = l \cdot w$ or $A = lw$

GUIDED PRACTICE

17.
$$
\begin{array}{r}
\overset{3}{15} \\
\times\ \ 7 \\
\hline
105
\end{array}
$$

19.
$$
\begin{array}{r}
\overset{3}{34} \\
\times\ \ 8 \\
\hline
272
\end{array}
$$

21. 100 has 2 zeros : attach 2 zeros : 3,700

23. 10 has 1 zero : attach 1 zero : 750

25. 10,000 has 4 zeros : attach 4 zeros : 1,070,000

27. 1,000 has 3 zeros : attach 3 zeros : 512,000

29. $68 \cdot 4 = 272$
$68 \cdot 40 = 2,720$

31. $56 \cdot 2 = 112$
$56 \cdot 200 = 11,200$

33. $13 \cdot 3 = 39$
$130(3,000) = 390,000$

35. $27 \cdot 4 = 108$
$2,700(40,000) = 108,000,000$

37.
$$\begin{array}{r} 128 \\ \times \quad 73 \\ \hline 384 \\ 8,960 \\ \hline 9,344 \end{array}$$

39.
$$\begin{array}{r} 287 \\ \times \quad 64 \\ \hline 1,148 \\ 17,220 \\ \hline 18,368 \end{array}$$

41. $602 \cdot 679$
$= 600 \cdot 679 + 2 \cdot 679$
$= 407,400 + 1,358$
$= 408,758$

43. $3,002(5,619)$
$= 3,000(5,619) + 2(5,619)$
$= 16,857,000 + 11,238$
$= 16,868,238$

45. $(18 \cdot 20) \cdot 5 = 18 \cdot (20 \cdot 5) = 18 \cdot 100 = 1,800$

47. $250 \cdot (4 \cdot 135) = (250 \cdot 4) \cdot 135 = 1000 \cdot 135 = 135,000$

49. $90 \cdot 200 = 18,000$

51. $200 \cdot 2,000 = 400,000$

53. $6 \cdot 14 = 84 \, in^2$

55. $12 \cdot 12 = 144 \, in^2$

TRY IT YOURSELF

57.
$$\begin{array}{r} \overset{2}{2}13 \\ \times \quad 7 \\ \hline 1,491 \end{array}$$

59.
$$\begin{array}{r} \overset{1}{3}4,474 \\ \times \quad 2 \\ \hline 68,948 \end{array}$$

61.
$$\begin{array}{r} 99 \\ \times \quad 77 \\ \hline 693 \\ 6,930 \\ \hline 7,623 \end{array}$$

63. $44(55)(0) = 0$

65. $53 \cdot 3 = 159$
$53 \cdot 30 = 1,590$

67.
$$\begin{array}{r} 754 \\ \times \quad 59 \\ \hline 6786 \\ +37700 \\ \hline 44,486 \end{array}$$

69.
$$\begin{array}{r} 2978 \\ \times 3004 \\ \hline 11\ 912 \\ 00\ 000 \\ 000\ 000 \\ +8\ 934\ 000 \\ \hline 8{,}945{,}912 \end{array}$$

71.
$$\begin{array}{r} 916 \\ \times\quad 409 \\ \hline 8\ 244 \\ 00\ 000 \\ +366\ 400 \\ \hline 374{,}644 \end{array}$$

73. $25 \cdot (4 \cdot 99) = (25 \cdot 4) \cdot 99 = 100 \cdot 99 = 9{,}900$

75. $48 \cdot 5 = 240$
$4{,}800 \cdot 500 = 2{,}400{,}000$

77.
$$\begin{array}{r} 2{,}779 \\ \times\quad 128 \\ \hline 22\ 232 \\ 55\ 580 \\ +277\ 900 \\ \hline 335{,}712 \end{array}$$

79.
$$\begin{array}{r} 370 \\ \times 450 \\ \hline 000 \\ 18\ 500 \\ +148\ 000 \\ \hline 166{,}500 \end{array}$$

APPLICATIONS

81. $2 \cdot 36 = 72$ cups of raisins

83. $12 \cdot 17 = 204$ grams of fat

85. $60 \cdot 65 = 3{,}900$ times per minute

87. $12 \cdot 5{,}280 = 63{,}360\, in$ in a mile

89. $250 \cdot 308 = 77{,}000$ words

91. $435 \cdot 169{,}300 = \$73{,}645{,}500$ per year

93. $8 \cdot 9 = 72$ entries

95. $17 \cdot 33 = 561$: There are 561 students and 1 instructor, so since $562 < 570$ they are O.K.

97. $3 \cdot 6 = 18$ hours asleep

99. $33 \cdot 42 = \$1{,}386$ per night

101. $2 \cdot 3 \cdot 14 = 6 \cdot 14 = 84$ pills

103. $3 \cdot 18 = 54\, ft^2$

105. Perimeter:
$360 + 270 + 360 + 270 = 1{,}260\, mi$

Area: $360 \cdot 270 = 97{,}200\, mi^2$

WRITING

107. 1 foot is a unit of length, while 1 square foot is a unit of area.

REVIEW

109. $10{,}357 + 9{,}809 + 476 = 20{,}642$

Section 1.5: Dividing Whole Numbers

VOCABULARY

1. $\underset{dividend}{12} \div \underset{divisor}{4} = \underset{quotient}{3}$

$divisor \rightarrow 4\overline{)12} \begin{array}{l}\leftarrow quotient \\ \leftarrow dividend\end{array}$

$\dfrac{dividend \rightarrow 12}{divisor \rightarrow 4} = 3 \leftarrow quotient$

3. The problem $6\overline{)246}$ is written in long division form.

5. One number is <u>divisible</u> by another number if, when we divide them, the remainder is 0.

CONCEPTS

7. a. 7 groups of 3

   b. 5 groups of 4, 2 left over

9. a. $\dfrac{25}{25} = 1$ b. $\dfrac{6}{1} = 6$

   c. $\dfrac{100}{0}$ is undefined   d. $\dfrac{0}{12} = 0$

11. a. $5\overline{)1147}$ quotient 2   b. $9\overline{)587}$ quotient 6

    c. $23\overline{)7501}$ quotient 3   d. $16\overline{)892}$ quotient 5

13. $\begin{array}{r} 37 \\ \times\ 9 \\ \hline 333 \end{array}$

15. a. 0 or 5

    b. 2 and 3

    c. sum

    d. 10

NOTATION

17. $\div$, $\overline{)}$, $-$

GUIDED PRACTICE

19. $9\overline{)45}$ quotient 5 because $5 \cdot 9 = 45$.

21. $44 \div 11 = 4$ because $4 \cdot 11 = 44$.

23. $7 \cdot 3 = 21$

25. $6 \cdot 12 = 72$

27. $\begin{array}{r} 16 \\ 6\overline{)96} \\ \underline{-6} \\ 36 \\ \underline{-36} \\ 0 \end{array}$

Check: $6(16) = 96$ ☺

29. $\begin{array}{r} 29 \\ 3\overline{)87} \\ \underline{-6} \\ 27 \\ \underline{-27} \\ 0 \end{array}$

Check: $3(29) = 87$ ☺

31. $\begin{array}{r} 325 \\ 7\overline{)2275} \\ \underline{-21} \\ 17 \\ \underline{-14} \\ 35 \\ \underline{-35} \\ 0 \end{array}$

Check: $7(325) = 2275$ ☺

33. $\begin{array}{r} 218 \\ 9\overline{)1962} \\ \underline{-18} \\ 16 \\ \underline{-9} \\ 72 \\ \underline{-72} \\ 0 \end{array}$

Check: $9(218) = 1962$ ☺

35.
$$\begin{array}{r} 504 \\ 62\overline{)31248} \\ -310 \\ \hline 24 \\ -0 \\ \hline 248 \\ -248 \\ \hline 0 \end{array}$$

Check: $62(504) = 31,248$ ☺

37.
$$\begin{array}{r} 602 \\ 37\overline{)22274} \\ -222 \\ \hline 07 \\ -0 \\ \hline 74 \\ -74 \\ \hline 0 \end{array}$$

Check: $37(602) = 22,274$ ☺

39.
$$\begin{array}{r} 39 \\ 24\overline{)951} \\ -72 \\ \hline 231 \\ -216 \\ \hline 15 \end{array}$$

39 R 15

Check: $39 \cdot 24 + 15 = 951$ ☺

41.
$$\begin{array}{r} 21 \\ 46\overline{)999} \\ -92 \\ \hline 79 \\ -46 \\ \hline 33 \end{array}$$

21 R 33

Check: $21 \cdot 46 + 33 = 999$ ☺

43.
$$\begin{array}{r} 47 \\ 524\overline{)24714} \\ -2096 \\ \hline 3754 \\ -3668 \\ \hline 86 \end{array}$$

Check: $47 \cdot 524 + 86 = 24,714$ ☺

45.
$$\begin{array}{r} 19 \\ 178\overline{)3514} \\ -178 \\ \hline 1734 \\ -1602 \\ \hline 132 \end{array}$$

Check: $19 \cdot 178 + 132 = 3,514$ ☺

	Divisible by	2	3	4	5	6	9	10
47.	2,940	Y	Y	Y	Y	Y		Y
49.	43,785		Y		Y		Y	
51.	181,223							
53.	9,499,200	Y	Y	Y	Y	Y		Y

55. 10 has 1 zero : take away 1 zero : 70

57. Begin by cancelling a zero from each.
$$\begin{array}{r} 22 \\ 45\overline{)990} \\ -90 \\ \hline 90 \\ -90 \\ \hline 0 \end{array}$$

59. $360,000 \div 40 = 9,000$

61. $50,000 \div 1,000 = 50$

TRY IT YOURSELF

63.
$$\begin{array}{r}4325\\6\overline{)25950}\\\underline{-24}\\19\\\underline{-18}\\15\\\underline{-12}\\30\\\underline{-30}\\0\end{array}$$

65.
$$\begin{array}{r}6\\9\overline{)54}\\\underline{-54}\\0\end{array}$$

67.
$$\begin{array}{r}8\\31\overline{)273}\\\underline{-248}\\25\end{array}$$

8 R 25

69. Begin by cancelling 2 zeros from each.

$$\begin{array}{r}160\\4\overline{)640}\\\underline{-4}\\24\\\underline{-24}\\0\end{array}$$

71.
$$\begin{array}{r}106\\7\overline{)745}\\\underline{-7}\\04\\\underline{-0}\\45\\\underline{-42}\\3\end{array}$$

106 R 3

73.
$$\begin{array}{r}509\\29\overline{)14761}\\\underline{-145}\\26\\\underline{-0}\\261\\\underline{-261}\\0\end{array}$$

75.
$$\begin{array}{r}3080\\175\overline{)539000}\\\underline{-525}\\140\\\underline{-0}\\1400\\\underline{-1400}\\0\end{array}$$

77.
$$\begin{array}{r}5\\15\overline{)75}\\\underline{-75}\\0\end{array}$$

79.
$$\begin{array}{r}23\\212\overline{)5087}\\\underline{-424}\\847\\\underline{-636}\\211\end{array}$$

23 R 211

81.
$$\begin{array}{r}30\\42\overline{)1273}\\\underline{-126}\\13\\\underline{-0}\\13\end{array}$$

30 R 13

83. 1,000 has 3 zeros : take away 3 zeros : 89

85. $$\begin{array}{r} 7 \\ 8\overline{)57} \\ \underline{-56} \\ 1 \end{array}$$

7 R 1

APPLICATIONS

87. $2500 \div 4 = 625$ tickets

89. $405 \div 15 = 27$ trips

91. $50 \div 23 = 2 R 4$ Each student got 2, with 4 left over.

93. $640 \div 68 = 9 R 28$ It can be filled 9 times with 28 oz. left

95. $58,000 \div 4 = 14,500$ There are 14,500 lbs. on each jack.

97. $25,200 \div 240 = \$105$ per book

99. $700 \div 140 = 5$ miles per gallon

101. $156 \div 12 = 13$ They should order 13 dozen donuts

103. $216 \div 7 \approx 30.86$ - teams won't have the same number

    $216 \div 8 = 27$ - not an even number of teams

    $216 \div 9 = 24$ GOOD CHOICE

    $216 \div 10 = 21.6$ - teams won't have the same number

    There are 24 teams with 9 girls each.

105. Divide each by 12: Nursing: $4,344 ; Marketing: $3,622 ; History: $2,996

WRITING

107. Find out how many times you must subtract 6 from 24 to get 0.

109. $30 - 2 \cdot 8 = 30 - 16 = 14$

    Since 14 is divisible by 7, 308 is also.

REVIEW

111. $2,903 + 378 = 3,281$

113. $2,903 \times 378 = 1,097,334$

Section 1.6: Problem Solving

VOCABULARY

1. A <u>strategy</u> is a careful plan or method.

3. subtraction

5. multiplication

7. addition

9. multiplication

11. division

CONCEPTS

13. Analyze, Form, Solve, State, Check

15. $15 \cdot 8 \div 3 = 120 \div 3 = 40$

GUIDED PRACTICE

17. $9 \cdot 21,605 = \$194,445$

19. $274 - 95 = 179$ *I Love Lucy* episodes

21. $98 \div 7 = 14$ servings of dark chocolate

23.

Act	Scenes
1	5
2	6
3	5
4	5
5	3
Total	24

24 scenes in total.

25. 5 lbs. = 80 ounces

$80 \div 3 = 26 \ R \ 2$

He can make 26 rolls, with 2 ounces of dough left over.

27. $81 - 26 + 13 = 55 + 13 = 68$ documents

TRY IT YOURSELF

29. Canada:
$3,287,243 - 2,342,949 = 944,294 \, mi^2$

US: $944,294 - 71,730 = 872,564 \, mi^2$

31. $998 + 411 + 337 + 372 + 267 + 238 = \$2,623$ million

33. $20,735 - 375 = 20,360$ submitted applications

35. $593 - 516 = \$77$ per person

$6 \cdot 77 = \$462$ for the family

37. 2 coats of $9,800 \, ft^2$ is $19,600 \, ft^2$

$19,600 \div 350 = 56$ gallons of paint

39. $27 + 14 + 13 = 54$ GB used

$80 - 54 = 26$ GB free

41. $379 + 47 = 426 \, ft.$ tall

43. $60 \cdot 24 \cdot 7 = 10,080$ minutes in a week

45. $516 \cdot 12 = 6,192$ bricks ordered

$6,192 \div 430 = 14 \ R \ 172$

14 fireplaces with 172 bricks to spare

47. $15 \cdot 15 = 225$ total squares

$225 - 46 = 179$ squares with letters

49. $1,989 + 125 + 296 - 1680 = \$730$

51. $3 \cdot 7,926 = 23,778 \, mi.$ of running

53. $59 + 26 + 23 + 37 + 11 = \$156$

Since $8 \cdot 20 = \$160$ he used 8 bills and received $160 - 156 = \$4$ in change.

55. $50 \cdot 2 + 4 \cdot 3 + 1 \cdot 1 = 100 + 12 + 1 = 113$ points

57. Whole garden: $27 \cdot 19 = 513 \, ft^2$

Arable space: $513 - 125 = 388 \, ft^2$

WRITING

59. The car dealership offered a sale price of $6,200 less than the suggested price of $25,500. Find the asking price of the car.

61. Six people recently split a lottery jackpot of $2,460,000. How much does each player receive?

REVIEW

63. Adding upwards gives 12,787 – there was an error.

65. Estimate: $70 \times 60 = 4,200$. The answer doesn't seem reasonable.

Section 1.7: Prime Factors and Exponents

VOCABULARY

1. Numbers that are multiplied together are called <u>factors</u>.

3. A <u>prime</u> number is a whole number greater than 1 that has only 1 and itself as factors.

5. To prime factor a number means to write it as a product of only <u>prime</u> numbers.

7. In the exponential expression 6^4, the number 6 is the <u>base</u> and 4 is the <u>exponent</u>.

CONCEPTS

9. $1 \cdot 45 = 45 \quad 3 \cdot 15 = 45 \quad 5 \cdot 9 = 45$

The factors of 45, in order from least to greatest, are 1, 3, 5, 9, 15, 45.

11. Yes

13. a. even, odd

    b. 0, 2, 4, 6, 8, 10, 12, 14, 16, 18

    c. 1, 3, 5, 7, 9, 11, 13, 15, 17, 19

15. The blank should be a 6.

    The prime factorization of 150 is $2 \cdot 3 \cdot 5 \cdot 5$.

17. 2$\vert$150
    3$\vert$75
    5$\vert$25
    5

    The prime factorization of 150 is $2 \cdot 3 \cdot 5 \cdot 5$.

NOTATION

19. a. base 7; exponent 6

    a. base 15; exponent 1

GUIDED PRACTICE

21. 1, 2, 5, 10

23. 1, 2, 4, 5, 8, 10, 20, 40

25. 1, 2, 3, 6, 9, 18

27. 1, 2, 4, 11, 22, 44

29. 1, 7, 11, 77

31. 1, 2, 4, 5, 10, 20, 25, 50, 100

33. $2 \cdot 4$

35. $3 \cdot 9$

37. $7 \cdot 7$

39. $2 \cdot 10$ or $4 \cdot 5$

41. $30 = 2 \cdot 15 = 2 \cdot 3 \cdot 5$

43. $63 = 3 \cdot 21 = 3 \cdot 3 \cdot 7$

45. $54 = 6 \cdot 9 = 2 \cdot 3 \cdot 9$ or $3 \cdot 3 \cdot 6$

47. $60 = 2 \cdot 3 \cdot 10 = 2 \cdot 5 \cdot 6 = 2 \cdot 2 \cdot 15 = 3 \cdot 4 \cdot 5$

49. 11 : 1 and 11

51. 37 : 1 and 37

53. Yes

55. No $(3 \cdot 3 \cdot 11)$

57. No $(3 \cdot 17)$

59. Yes

61. $30 = 6 \cdot 5 = 2 \cdot 3 \cdot 5$

63. $39 = 3 \cdot 13$

65. $99 = 9 \cdot 11 = 3 \cdot 3 \cdot 11 = 3^2 \cdot 11$

67. $162 = 2 \cdot 81 = 2 \cdot 9 \cdot 9 = 2 \cdot 3 \cdot 3 \cdot 3 \cdot 3 = 2 \cdot 3^4$

69. $64 = 8 \cdot 8 = 2 \cdot 4 \cdot 2 \cdot 4 = 2 \cdot 2 \cdot 2 \cdot 2 \cdot 2 \cdot 2 = 2^6$

71. $147 = 3 \cdot 49 = 3 \cdot 7 \cdot 7 = 3 \cdot 7^2$

73. $220 = 22 \cdot 10 = 2 \cdot 11 \cdot 2 \cdot 5 = 2^2 \cdot 5 \cdot 11$

75. $102 = 2 \cdot 51 = 2 \cdot 3 \cdot 17$

77. $2 \cdot 2 \cdot 2 \cdot 2 \cdot 2 = 2^5$

79. $5 \cdot 5 \cdot 5 \cdot 5 = 5^4$

81. $4(4)(8)(8)(8) = 4^2 (8^3)$

83. $7 \cdot 7 \cdot 7 \cdot 9 \cdot 9 \cdot 7 \cdot 7 \cdot 7 \cdot 7 = 7^7 \cdot 9^2$

85. a. $3^4 = 3 \cdot 3 \cdot 3 \cdot 3 = 81$ b. $4^3 = 4 \cdot 4 \cdot 4 = 64$

87. a. $2^5 = 2 \cdot 2 \cdot 2 \cdot 2 \cdot 2 = 32$ b. $5^2 = 5 \cdot 5 = 25$

89. a. $7^3 = 7 \cdot 7 \cdot 7 = 343$ b. $3^7 = 3 \cdot 3 \cdot 3 \cdot 3 \cdot 3 \cdot 3 \cdot 3 = 2,187$

91. a. $9^1 = 9$ b. $1^9 = 1$

93. $2 \cdot 3 \cdot 3 \cdot 5 = 90$

95. $7 \cdot 11^2 = 7 \cdot 121 = 847$

97. $3^2 \cdot 5^2 = 9 \cdot 25 = 225$

99. $2^3 \cdot 3^3 \cdot 13 = 8 \cdot 27 \cdot 13 = 2,808$

APPLICATIONS

101. Factors of 28: 1, 2, 4, 7, 14, 28

$1 + 2 + 4 + 7 + 14 = 28$

103. 2^2 square units, 3^2 square units, 4^2 square units

WRITING

105. Multiply the factors together to verify you get the original number

107. $1^2 = 1^3 = 1^4 = 1$. Any power of 1 is 1.

REVIEW

109. $8 \cdot 15 + 5 = 120 + 5 = 125$ band members

Section 1.8: The LCM and the GCF

VOCABULARY

1. The <u>multiples</u> of a number are the products of that number and 1, 2, 3, 4, 5, and so on.

3. One number is <u>divisible</u> by another number if, when dividing them, we get a remainder of 0.

CONCEPTS

5. a. 12

b. In general, the LCM of two whole numbers is the <u>smallest</u> whole number that is divisible by both numbers.

7. a. 20

b. 20

9. a. 2 appears twice with 36.

b. 3 appears twice with 90.

c. 5 appears once with 90.

d. LCM = $2 \cdot 2 \cdot 3 \cdot 3 \cdot 5 = 180$

11. a. 2 appears twice with 12.

b. 3 appears three times with 54.

c. LCM = $2^2 \cdot 3^3 = 108$

13. a. 2, 3, and 5 are common to both.

b. GCF = $2 \cdot 3 \cdot 5 = 30$

NOTATION

15. a. The abbreviation for the greatest common factor is <u>GCF</u>.

b. The abbreviation for the least common multiple is <u>LCM</u>.

GUIDED PRACTICE

17. 4, 8, 12, 16, 20, 24, 28, 32

19. 11, 22, 33, 44, 55, 66, 77, 88

21. 8, 16, 24, 32, 40, 48, 56, 64

23. 20, 40, 60, 80, 100, 120, 140, 160

25. 5 is not divisible by 3.

10 is not divisible by 3.

15 is divisible by 3.

LCM(3,5) = 15

27. 12 is not divisible by 8.

24 is divisible by 8.

LCM(8,12) = 24

29. 11 is not divisible by 5.

22 is not divisible by 5.

33 is not divisible by 5.

44 is not divisible by 5.

55 is divisible by 5.

LCM(5,11) = 55

31. 7 is not divisible by 4.

14 is not divisible by 4.

21 is not divisible by 4.

28 is divisible by 4.

LCM(4,7) = 28

33. 6 is not divisible by 3 and 4.

12 is divisible by 3 and 4.

LCM(3,4,6) = 12.

35. 10 is not divisible by 2 and 3.

20 is not divisible by 2 and 3.

30 is divisible by 2 and 3.

LCM(2,3,10) = 30

37. $16 = 2^4$
$20 = 2^2 \cdot 5$
$LCM = 2^4 \cdot 5 = 16 \cdot 5 = 80$

39. $30 = 2 \cdot 3 \cdot 5$
$50 = 2 \cdot 5^2$
$LCM = 2 \cdot 3 \cdot 5^2 = 6 \cdot 25 = 150$

41. $35 = 5 \cdot 7$
$45 = 3^2 \cdot 5$
$LCM = 3^2 \cdot 5 \cdot 7 = 9 \cdot 35 = 315$

43. $100 = 2^2 \cdot 5^2$
$120 = 2^3 \cdot 3 \cdot 5$
$LCM = 2^3 \cdot 3 \cdot 5^2 = 600$

45. $6 = 2 \cdot 3$
$24 = 2^3 \cdot 3$
$36 = 2^2 \cdot 3^2$
$LCM = 2^3 \cdot 3^2 = 72$

47. $5 = 5$
$12 = 2^2 \cdot 3$
$15 = 3 \cdot 5$
$LCM = 2^2 \cdot 3 \cdot 5 = 60$

49. $4 = \underline{2} \cdot 2$
$6 = \underline{2} \cdot 3$
$GCF = 2$

51. $9 = 3 \cdot \underline{3}$
$12 = 2 \cdot 2 \cdot \underline{3}$
$GCF = 3$

53. $22 = 2 \cdot \underline{11}$
$33 = 3 \cdot \underline{11}$
$GCF = 11$

55. $15 = \underline{3} \cdot \underline{5}$
$30 = 2 \cdot \underline{3} \cdot \underline{5}$
$GCF = 3 \cdot 5 = 15$

57. $18 = \underline{2} \cdot \underline{3} \cdot 3$
$96 = \underline{2} \cdot 2 \cdot 2 \cdot 2 \cdot 2 \cdot \underline{3}$
$GCF = 2 \cdot 3 = 6$

59. $28 = \underline{2} \cdot 2 \cdot \underline{7}$
$42 = \underline{2} \cdot 3 \cdot \underline{7}$
$GCF = 2 \cdot 7 = 14$

61. $16 = 2 \cdot 2 \cdot 2 \cdot 2$
$51 = 3 \cdot 17$
$GCF = 1$

63. $81 = 3 \cdot 3 \cdot 3 \cdot 3$
$125 = 5 \cdot 5 \cdot 5$
$GCF = 1$

65. $12 = 2 \cdot 2 \cdot 3$
$68 = 2 \cdot 2 \cdot 17$
$92 = 2 \cdot 2 \cdot 23$
$GCF = 2 \cdot 2 = 4$

67. $72 = 2 \cdot 2 \cdot 2 \cdot 3 \cdot 3$
$108 = 2 \cdot 2 \cdot 3 \cdot 3 \cdot 3$
$144 = 2 \cdot 2 \cdot 2 \cdot 2 \cdot 3 \cdot 3$
$GCF = 2 \cdot 2 \cdot 3 \cdot 3 = 36$

TRY IT YOURSELF

69. $100 = 2 \cdot 2 \cdot 5 \cdot 5$
$120 = 2 \cdot 2 \cdot 2 \cdot 3 \cdot 5$
$LCM = 2 \cdot 2 \cdot 2 \cdot 3 \cdot 5 \cdot 5 = 600$
$GCF = 2 \cdot 2 \cdot 5 = 20$

71. $14 = 2 \cdot 7$
$140 = 2 \cdot 2 \cdot 5 \cdot 7$
$LCM = 2 \cdot 2 \cdot 5 \cdot 7 = 140$
$GCF = 2 \cdot 7 = 14$

73. $66 = 2 \cdot 3 \cdot 11$
$198 = 2 \cdot 3 \cdot 3 \cdot 11$
$242 = 2 \cdot 11 \cdot 11$
$LCM = 2 \cdot 3 \cdot 3 \cdot 11 \cdot 11 = 2,178$
$GCF = 2 \cdot 11 = 22$

75. $8 = 2 \cdot 2 \cdot 2$
$9 = 3 \cdot 3$
$49 = 7 \cdot 7$
$LCM = 2 \cdot 2 \cdot 2 \cdot 3 \cdot 3 \cdot 7 \cdot 7 = 3,538$
$GCF = 1$

77. $120 = 2 \cdot 2 \cdot 2 \cdot 3 \cdot 5$
$125 = 5 \cdot 5 \cdot 5$
$LCM = 2 \cdot 2 \cdot 2 \cdot 3 \cdot 5 \cdot 5 \cdot 5 = 3,000$
$GCF = 5$

79. $34 = 2 \cdot 17$
$68 = 2 \cdot 2 \cdot 17$
$102 = 2 \cdot 3 \cdot 17$
$LCM = 2 \cdot 2 \cdot 3 \cdot 17 = 204$
$GCF = 2 \cdot 17 = 34$

81. $46 = 2 \cdot 23$
$69 = 3 \cdot 23$
$LCM = 2 \cdot 3 \cdot 23 = 138$
$GCF = 23$

83. $50 = 2 \cdot 5 \cdot 5$
$81 = 3 \cdot 3 \cdot 3 \cdot 3$
$LCM = 2 \cdot 3 \cdot 3 \cdot 3 \cdot 3 \cdot 5 \cdot 5 = 4,050$
$GCF = 1$

APPLICATIONS

85.

1st	2nd	3rd
7,500mi	15,000mi	22,500mi
4th	5th	6th
30,000mi	37,500mi	45,000mi

87. LCM(45,60) = 180 minutes, or 3 hours

89. LCM of 10 and 12 = 60

5 packs of buns, 6 packs of hot dogs

91. LCM(6,8) = 24

4 sheets wide by 3 sheets tall = 12 sheets

93. a. GCF(28,21,63) = 7

The most that the art supplies cost a student is $7.

b. 4 students, 3 students, 9 students

WRITING

95. Find the prime factorization of both 8 and 28, then take each factor present the largest number of times it appears.

97. Since each factor has only one 3, the LCM only needs one 3.

REVIEW

99. $9,999 + 1,111 = 11,110$

101. $305 \cdot 50 = 15,250$

Section 1.9: Order of Operations

VOCABULARY

1. Numbers are combined with the operations of addition, subtraction, multiplication, and division to create <u>expressions</u>.

3. The grouping symbols ( ) are called <u>parentheses</u>, and the symbols [ ] are called <u>brackets</u>.

5. In the expression $9+6[8+6(4-1)]$, the parentheses are the <u>inner</u> most grouping symbols and the brackets are the <u>outer</u> most grouping symbols.

CONCEPTS

7. a. $5(2)^2 - 1$: square, multiply, subtract

b. $15 + 90 - (2 \cdot 2)^3$: multiply, cube, add, subtract

c. $7 \cdot 4^2$: square, multiply

d. $(7 \cdot 4)^2$: multiply, square

9. multiply ; square

NOTATION

11. The fraction bar groups the numerator and denominator.

13. We read the expression $16 - (4 + 9)$ as "16 minus the <u>quantity</u> of 4 plus 9."

15. $7 \cdot 4 - 5(2)^2 = 7 \cdot 4 - 5(4)$
$= 28 - 20$
$= 8$

17. $[4(2+7)] - 4^2 = [4(9)] - 4^2$
$= 36 - 4^2$
$= 36 - 16$
$= 20$

GUIDED PRACTICE

19. $3 \cdot 5^2 - 28$
$= 3 \cdot 25 - 28$
$= 75 - 28$
$= 47$

21. $6 \cdot 3^2 - 41$
$= 6 \cdot 9 - 41$
$= 54 - 41$
$= 13$

23. $52 - 6 \cdot 3 + 4$
$= 52 - 18 + 4$
$= 34 + 4$
$= 38$

25. $32 - 9 \cdot 3 + 31$
$= 32 - 27 + 31$
$= 5 + 31$
$= 36$

27. $192 \div 4 - 4(2)3$
$= 48 - 24$
$= 24$

29. $252 \div 3 - 6(2)6$
$= 84 - 72$
$= 12$

31. a. $26 - 2 + 9 = 24 + 9 = 33$

b. $26 - (2 + 9) = 26 - 11 = 15$

33. a. $51 - 16 + 8 = 35 + 8 = 43$

b. $51 - (16 + 8) = 51 - 24 = 27$

35. $(4+6)^2 = 10^2 = 100$

37. $(3+5)^3 = 8^3 = 512$

39. $8 + 4(29 - 5 \cdot 3)$
$= 8 + 4(29 - 15)$
$= 8 + 4(14)$
$= 8 + 56$
$= 64$

41. $77 + 9(38 - 4 \cdot 6)$
$= 77 + 9(38 - 24)$
$= 77 + 9(14)$
$= 77 + 126$
$= 203$

43.
$$46 + 3\left[5^2 - 4(9-5)\right]$$
$$= 46 + 3\left[25 - 4(4)\right]$$
$$= 46 + 3\left[25 - 16\right]$$
$$= 46 + 3[9]$$
$$= 46 + 27$$
$$= 73$$

45.
$$81 + 9\left[7^2 - 7(11-4)\right]$$
$$= 81 + 9\left[49 - 7(7)\right]$$
$$= 81 + 9[49 - 9]$$
$$= 81 + 9[0]$$
$$= 81$$

47.
$$\frac{2(50) - 4}{2(4^2)}$$
$$= \frac{100 - 4}{2 \cdot 16}$$
$$= \frac{96}{32}$$
$$= 3$$

49.
$$\frac{25(8) - 8}{6(2^3)}$$
$$= \frac{200 - 8}{6 \cdot 8}$$
$$= \frac{192}{48}$$
$$= 4$$

51. $\dfrac{6+9+4+3+8}{5} = \dfrac{30}{5} = 6$

53. $\dfrac{3+5+9+1+7+5}{6} = \dfrac{30}{6} = 5$

55. $\dfrac{19+15+17+13}{4} = \dfrac{64}{4} = 16$

57. $\dfrac{5+8+7+0+3+1}{6} = \dfrac{24}{6} = 4$

TRY IT YOURSELF

59. $(8-6)^2 + (4-3)^2 = 2^2 + 1^2$
$= 4 + 1 = 5$

61. $2 \cdot 3^4 = 2 \cdot 81 = 162$

63. $7 + 4 \cdot 5 = 7 + 20 = 27$

65. $(7-4)^2 + 1 = 3^2 + 1 = 9 + 1 = 10$

67. $\dfrac{10+5}{52-47} = \dfrac{15}{5} = 3$

69.
$$5 \cdot 10^3 + 2 \cdot 10^2 + 3 \cdot 10^1 + 9$$
$$= 5 \cdot 1000 + 2 \cdot 100 + 3 \cdot 10 + 9$$
$$= 5,000 + 200 + 30 + 9$$
$$= 5,239$$

71. $20 - 10 + 5 = 10 + 5 = 15$

73. $25 \div 5 \cdot 5 = 5 \cdot 5 = 25$

75.
$$150 - 2(2 \cdot 6 - 4)^2$$
$$= 150 - 2(12 - 4)^2$$
$$= 150 - 2 \cdot 8^2$$
$$= 150 - 2 \cdot 64$$
$$= 150 - 128$$
$$= 22$$

77.
$$190 - 2\left[10^2 - (5 + 2^2)\right] + 45$$
$$= 190 - 2\left[100 - (5+4)\right] + 45$$
$$= 190 - 2[100 - 9] + 45$$
$$= 190 - 2[91] + 45$$
$$= 190 - 182 + 45$$
$$= 53$$

79. $2 + 3(0) = 2$

81.
$$\frac{(5-3)^2+2}{4^2-(8+2)}$$
$$=\frac{2^2+2}{4^2-10}$$
$$=\frac{4+2}{16-10}$$
$$=\frac{6}{6}$$
$$=1$$

83. $4^2+3^2=16+9=25$

85. $3+2\cdot 3^4\cdot 5=3+2\cdot 81\cdot 5$
$=3+810=813$

87. $60-\left(6+\frac{40}{2^3}\right)$
$=60-\left(6+\frac{40}{8}\right)$
$=60-(6+5)$
$=60-11$
$=49$

89.
$$\frac{(3+5)^2+2}{2(8-5)}$$
$$=\frac{8^2+2}{2(3)}$$
$$=\frac{64+2}{6}$$
$$=\frac{66}{6}$$
$$=11$$

91. $(18-12)^3-5^2=6^3-5^2=216-25=191$

93. $30(1)^2-4(2)+12=30-8+12=34$

95. $16^2-\frac{25}{5}+6(3)4$
$=256-5+72$
$=323$

97. $\frac{3^2-2^2}{(3-2)^2}=\frac{9-4}{1^2}=\frac{5}{1}=5$

99. $3\left(\frac{18}{3}\right)-2(2)=18-4=14$

101. $4\left[50-(3^3-5^2)\right]$
$=4\left[50-(27-25)\right]$
$=4[50-2]$
$=4[48]$
$=192$

103. $80-2\left[12-(5+4)\right]$
$=80-2[12-9]$
$=80-2[3]$
$=80-6$
$=74$

APPLICATIONS

105. $3\cdot 7+4\cdot 4+2\cdot 3$
$=21+16+6=\$43$

107. $3(8+7+8+8+7)$
$=3(38)=114$

109. brick: $3\cdot 3+1+1+3+3\cdot 5=29$

aphid: $3\left[1+2(3)+4+1+2\right]=42$

111. $2^2+3^2+5^2+7^2=4+9+25+49=87$

113. $\frac{75+80+83+80+77+72+86}{7}$
$=\frac{553}{7}$
$=79°$

115. $\frac{39+40+\cdots+42}{12}$
$=\frac{372}{12}$
$=31$ therms

117. $\dfrac{230+280+\cdots 375}{8}$

$= \dfrac{2400}{8}$

$= 300 \text{ calories}$

119. a. $1+4+35+85 = 125$

b. $1 \cdot 2{,}500 + 4 \cdot 500 + 35 \cdot 150 + 85 \cdot 25$
$= 2{,}500 + 2{,}000 + 5{,}250 + 2{,}125$
$= 11{,}875$

c. $\dfrac{11{,}875}{125} = \$95$

WRITING

121. Order of operations is necessary so that different people don't come up with different answers to the same question.

123. The multiplication of 2 and 3 takes precedence over the addition.

REVIEW

125. Two hundred fifty-four thousand, three hundred nine

Chapter 1 Review

1. 6 is in the ten thousands column.

3. 1 is the 1 billion place value.

5. a. ninety-seven thousand, two hundred eighty-three

b. five billion, four hundred forty-four million, sixty thousand, seventeen

7. 500,000+70,000+300+2

9.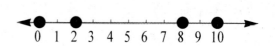

11. $9 > 7$

13. a. 2,507,300

b. 2,510,000

c. 2,507,350

d. 3,000,000

15. a.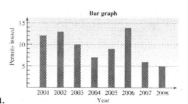

b.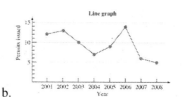

17. $\overset{1}{4}36$
$\underline{+27}$
463

19. $4+(36+19)$
$= 4+(55)$
$= 59$

21. $\overset{1}{5}{,}\overset{1}{3}\overset{1}{4}5$
$\underline{+\ \ 655}$
$6{,}000$

23. $\overset{1\ 2\ 2}{}$
 4,447
 7,478
 + 676
 ────────
 12,601

25. $\overset{\ \ 2\ 2}{}$
 226
 345
 859
 +1,291
 ──────
 2,721

It is not correct.

27. $600 + 800 + 10,000 + 40,000 + 8,000$
 $= 1,400 + 50,000 + 8,000$
 $= 51,400 + 8,000$
 $= 59,400$

29. $\overset{1\ 1\ \ 2\ 2\ 1\ \ 2\ 1}{}$
 89,379,287
 76,177,855
 +61,896,075
 ──────────
 227,453,217

31. $\overset{1\ 1}{}$
 717,900,000
 +606,800,000
 ────────────
 $1,324,700,000

33. 148
 −87
 ───
 61

35. $\overset{\ \ \ \ \ 2\ 15}{10{,}4\cancel{3}\cancel{5}}$
 −10,218
 ──────
 217

37. $750 − 259 + 14$
 $= 491 + 14$
 $= 505$

39. $\overset{1\ 1\ 1}{}$
 1,168
 +6,949
 ──────
 8,117

The subtraction is incorrect.

41. $200,000 − 40,000 = 160,000$

43. $12,975 − 3,800 + 4,270$
 $= 9,175 + 4,270$
 $= \$13,445$

45. $\overset{6}{}$
 47
 ×9
 ───
 423

47. $72 \cdot 10,000$: Since there are 4 zeros, move the decimal place 4 units to the right: 720,000

49. 157
 ×59
 ─────
 1,413
 +7,850
 ──────
 9,263

51. 5,624
 ×281
 ──────
 5,624
 449,920
 +1,124,800
 ──────────
 1,580,344

53. $7,000 \cdot 400 = 2,800,000$

55. a. $8 \cdot 0 = 0$ b. $7 \cdot 1 = 7$

57. $A = l \cdot w$
 $A = 8 \cdot 4$
 $A = 32 cm^2$

59. a. $365 \cdot 7 = 2,555 hr$

 b. $365 \cdot 9 = 3,285 hr$

61. Sarah: $12 \cdot 9 = \$108$

Santiago: $14 \cdot 8 = \$112$

Santiago earned more money.

63. $\dfrac{72}{4} = \dfrac{\cancel{4} \cdot 18}{\cancel{4}} = 18$

65.
$$\begin{array}{r} 37 \\ 39\overline{)1443} \\ -117 \\ \hline 273 \\ -273 \\ \hline 0 \end{array}$$

67.
$$\begin{array}{r} 23 \\ 54\overline{)1269} \\ -108 \\ \hline 189 \\ -162 \\ \hline 27 \end{array}$$
$23 R 27$

69. $\dfrac{0}{10} = 0$

71.
$$\begin{array}{r} 42 \\ 127\overline{)5347} \\ -508 \\ \hline 267 \\ -254 \\ \hline 13 \end{array}$$
$42 R 13$

73. $40 \cdot 4 = 160$

75. 364,545 is divisible by 3, 5, and 9.

77.
$$\begin{array}{r} 16 \\ 45\overline{)745} \\ -45 \\ \hline 295 \\ -270 \\ \hline 25 \end{array}$$

Each child will get 16 candies, with 25 left over.

79. $130° + 20° + 20° + 15°$
$= 150° + 20° + 15°$
$= 170° + 15°$
$= 185°F$

81. $12 \cdot 75 = 900 lb$

83. $15{,}000 \div 6 = 2{,}500$ boxes per day

85. $350 - 124 - 79$
$= 226 - 79$
$= 147$

147 of them were cattle.

87. $1, 2, 3, 6, 9, 18$

89. $20 = 2 \cdot 10 = 4 \cdot 5$

91. a. prime

b. composite

c. neither

d. neither

e. composite

f. prime

93. $42 = 2 \cdot 21 = 2 \cdot 3 \cdot 7$

95. $220 = 10 \cdot 22 = 2 \cdot 5 \cdot 2 \cdot 11 = 2^2 \cdot 5 \cdot 11$

97. $6 \cdot 6 \cdot 6 \cdot 6 = 6^4$

99. $5^3 = 5 \cdot 5 \cdot 5 = 25 \cdot 5 = 125$

101. $2^4 \cdot 7^2$
$= 2 \cdot 2 \cdot 2 \cdot 2 \cdot 7 \cdot 7$
$= 4 \cdot 4 \cdot 49$
$= 16 \cdot 49$
$= 784$

103. $9, 18, 27, 36, 45, 54, 63, 72, 81, 90$

105. $4 = 2 \cdot 2$
$6 = 2 \cdot 2 \cdot 3$
$LCM(4, 6) = 2 \cdot 2 \cdot 3 = 12$

107. $9 = 3 \cdot 3$
$15 = 3 \cdot 5$
$LCM(9, 15) = 3 \cdot 3 \cdot 5 = 45$

109. $18 = 2 \cdot 3 \cdot 3$
$21 = 3 \cdot 7$
$LCM(18, 21) = 2 \cdot 3 \cdot 3 \cdot 7 = 126$

111. $4 = 2 \cdot 2$
$14 = 2 \cdot 7$
$20 = 2 \cdot 2 \cdot 5$
$LCM(4, 14, 20) = 2 \cdot 2 \cdot 5 \cdot 7 = 140$

113. $8 = 2 \cdot 2 \cdot 2$
$12 = 2 \cdot 2 \cdot 3$
$GCF(8, 12) = 2 \cdot 2 = 4$

115. $30 = 2 \cdot 3 \cdot 5$
$40 = 2 \cdot 2 \cdot 2 \cdot 5$
$GCF(30, 40) = 2 \cdot 5 = 10$

117. $63 = 3 \cdot 3 \cdot 7$
$84 = 2 \cdot 2 \cdot 3 \cdot 7$
$GCF(63, 84) = 3 \cdot 7 = 21$

119. $48 = 2 \cdot 2 \cdot 2 \cdot 2 \cdot 3$
$72 = 2 \cdot 2 \cdot 2 \cdot 3 \cdot 3$
$120 = 2 \cdot 2 \cdot 2 \cdot 3 \cdot 5$
$GCF(48, 72, 120) = 2 \cdot 2 \cdot 2 \cdot 3 = 24$

121. $14 = 2 \cdot 7$
$21 = 3 \cdot 7$
$LCM(14, 21) = 2 \cdot 3 \cdot 7 = 42$

They will meet on the same day 42 days later.

123. $3^2 + 12 \cdot 3$
$= 9 + 36$
$= 45$

125. $(6 \div 2 \cdot 3)^2 \cdot 3$
$= (3 \cdot 3)^2 \cdot 3$
$= 9^2 \cdot 3$
$= 81 \cdot 3$
$= 243$

127. $2^3 \cdot 5 - 4 \div 2 \cdot 4$
$= 8 \cdot 5 - 2 \cdot 4$
$= 40 - 8$
$= 32$

129. $2 + 3\left(\dfrac{100}{10} - 2^2 \cdot 2\right)$
$= 2 + 3(10 - 4 \cdot 2)$
$= 2 + 3(10 - 8)$
$= 2 + 3(2)$
$= 2 + 6$
$= 8$

131. $\dfrac{4(6) - 6}{2(3^2)}$
$= \dfrac{24 - 6}{2(9)}$
$= \dfrac{18}{18}$
$= 1$

133.
$$7+3\left[3^3-10(4-2)\right]$$
$$=7+3\left[27-10(2)\right]$$
$$=7+3\left[27-20\right]$$
$$=7+3\left[7\right]$$
$$=7+21$$
$$=28$$

135.
$$\frac{80+74+66+88}{4}$$
$$=\frac{308}{4}$$
$$=77$$

Chapter 1 Test

1. a. The set of <u>whole</u> numbers is $\{0,1,2,3,4,5,\ldots\}$.

b. The symbols > and < are <u>inequality</u> symbols.

c. To *evaluate* an expression such as $58-33+9$ means to find its <u>value</u>.

d. The <u>area</u> of a rectangle is a measure of the amount of surface it encloses.

e. One number is <u>divisible</u> by another number if, when we divide them, the remainder is 0.

f. The grouping symbols ( ) are called <u>parentheses</u>, and the symbols [ ] are called <u>brackets</u>.

g. A <u>prime</u> number is a whole number greater than 1 that has only 1 and itself as factors.

3. a. 1 hundred

b. 0

5. a. $15>10$ b. $1,247<1,427$

7.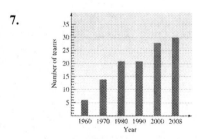

9.
```
  1 1 1 11
  136,231
   82,574
 +  6,359
 ────────
  225,164
```

11.
```
    2
   53
 ×  8
 ────
  424
```

13.
```
       72
    ─────
  6)432
    -42
    ───
     12
    -12
    ───
      0
```

15. $23\cdot 6=138$, now add 5 zeros: $13,800,000$

17. $50,000-7,000=43,000$

19. $23\cdot 23=529\,in^2$

21. $1260=10\cdot 126$
$=2\cdot 5\cdot 9\cdot 14$
$=2\cdot 5\cdot 3\cdot 3\cdot 2\cdot 7$
$=2^2\cdot 3^2\cdot 5\cdot 7$

23. $10,000-5,067=4,933$ tails

25. $12,255\div 3=4,085\,ft^2$

27. $1,350,000-26,000=1,324,000$
$1,324,000\div 4=\$331,000$

29. a. 0

b. 0

c. 1

d. undefined

31. $8 = 2 \cdot 2 \cdot 2$
$9 = 3 \cdot 3$
$12 = 2 \cdot 2 \cdot 3$
$LCM(8, 9, 12) = 2 \cdot 2 \cdot 2 \cdot 3 \cdot 3 = 72$

33. $24 = 2 \cdot 2 \cdot 2 \cdot 3$
$28 = 2 \cdot 2 \cdot 7$
$36 = 2 \cdot 2 \cdot 3 \cdot 3$
$GCF(24, 28, 36) = 2 \cdot 2 = 4$

35. It is divisible by 2, 3, 4, 5, 6, and 10.

37. $9 + 4 \cdot 5$
$= 9 + 20$
$= 29$

39. $20 + 2\left[4^2 - 2\left(6 - 2^2\right)\right]$
$= 20 + 2\left[16 - 2(6 - 4)\right]$
$= 20 + 2\left[16 - 2(2)\right]$
$= 20 + 2[16 - 4]$
$= 20 + 2[12]$
$= 20 + 24$
$= 44$

CHAPTER 2 The Integers

Section 2.1: An Introduction to the Integers

VOCABULARY

1. <u>Positive</u> numbers are greater than 0 and <u>negative</u> numbers are less than 0.

3. To <u>graph</u> an integer means to locate it on the number line and highlight it with a dot.

5. The <u>absolute value</u> of a number is the distance between the number and 0 on the number line.

CONCEPTS

7. a. -$225

   b. -10 sec.

   c. -3°

   d. -$12,000

   e. -1 mi.

9. a. The spacing is not uniform.

   b. The numbering scale is not uniform.

   c. The 0 is missing.

   d. The arrowheads aren't drawn.

11. a. -4 is 3 units right of -7.

    b. -2 is 4 units left of 2.

13. a. -7 is closer to -3 than 2.

    b. 8 is further from 1 than -5.

15. a. $15 > -12$

    b. $-5 < -4$

17.
Number	Opposite	Absolute value
-25	25	25
39	-39	39
0	0	0

NOTATION

19. a. $-(-8)$

    b. $|-8|$

    c. $8 - 8$

    d. $-|-8|$

21. a. We read $\geq$ as "is <u>greater</u> than or <u>equal</u> to."

    b. We read $\leq$ as "is <u>less</u> than or <u>equal</u> to."

GUIDED PRACTICE

23.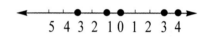

25.

27.

29.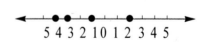

31. $-5 < 5$

33. $-12 < -6$

35. $-10 > -17$

37. $-325 > -532$

39. True

41. True

43. False

45. False

47. $|9| = 9$

49. $|-8| = 8$

51. $|-14| = 14$

53. $|180| = 180$

55. $-(-11) = 11$

57. $-(-4) = 4$

59. $-(-102) = 102$

61. $-(-561) = 561$

63. $-|20| = -20$

65. $-|6| = -6$

67. $-|-253| = -253$

69. $-|-0| = 0$

TRY IT YOURSELF

71. $|-12| \square -(-7)$
    $12 > 7$

73. $-|-71| \square -|-65|$
    $-71 < -65$

75. $-(-343) \square -(-161)$
    $343 > 161$

77. $-|-30| \square -|-(-8)|$
    $-30 < -8$

79. $-52, -22, -12, 12, 52, 82$

81. $5, 3, 1, -1, \boxed{-3}, \boxed{-5}, \boxed{-7}, \ldots$

APPLICATIONS

83. -31 lengths

85.
Time	Position
0 sec	0
1 sec	20
2 sec	5
3 sec	-40
4 sec	-120

87. Peaks: 2, 4 0 ; Valleys: -3, -5, -2

89. a. -1 : 1 below par

    b. -3 : 3 below par

    c. Most of the scores are below par.

91. a. $-20°$ to $-10°$

    b. $40°$

    c. $10°$

93. a. 200 years

    b. A.D.

    c. B.C.

    d. The birth of Christ

95. [Line graph showing temperature (Fahrenheit) across Mon, Tue, Wed, Thu, Fri]

WRITING

97. The opposite of a number is the number that is the same distance from 0 on the number line, but on the other side of 0.

99. Absolute value represents a distance, which cannot be negative.

101. Positive buoyancy would mean you float, negative would mean you sink. Neutral buoyancy would mean that you remain stationary in the water.

103. No it is not: 4 > 3 but -4 < -3.

REVIEW

105. 23,500

107. $2,842 - 2,081 = 761$

109. Associative Property of Multiplication

Section 2.2: Adding Integers

VOCABULARY

1. Two negative integers, as well as two positive integers, are said to have the same or <u>like</u> signs.

3. When 0 is added to a number, the number remains the same. We call 0 the additive <u>identity</u>.

5. <u>Commutative</u> property of addition: The order in which integers are added does not change their sum.

6. <u>Associative</u> property of addition: The way in which integers are grouped does not change their sum.

CONCEPTS

7. a. $|10| = 10$; $|-12| = 12$

b. -12

c. $|-12| - |10| = 12 - 10 = 2$

9. To add two integers with unlike signs, <u>subtract</u> their absolute values, the smaller from the larger. Then attach to that result the sign of the number with the <u>larger</u> absolute value.

11. a. yes

b. yes

c. no

d. no

13. a. 0

b. 0

NOTATION

15. $-16 + (-2) + (-1) = -18 + (-1)$
$= -19$

17. $(-3 + 8) + (-3) = 5 + (-3)$
$= 2$

GUIDED PRACTICE

19. Add the absolute values. Since both numbers are negative, the result is negative.

$6 + 3 = 9$ so $-6 + (-3) = -9$

21. Add the absolute values. Since both numbers are negative, the result is negative.

$5 + 5 = 10$ so $-5 + (-5) = -10$

23. Add the absolute values. Since both numbers are negative, the result is negative.

$51 + 11 = 62$ so $-51 + (-11) = -62$

25. Add the absolute values. Since both numbers are negative, the result is negative.

$69 + 27 = 96$ so $-69 + (-27) = -96$

27. Add the absolute values. Since both numbers are negative, the result is negative.

$248 + 131 = 379$ so $-248 + (-131) = -379$

29. Add the absolute values. Since both numbers are negative, the result is negative.

$565 + 309 = 874$ so $-565 + (-309) = -874$

31. Subtract the smaller absolute value from the larger. Since the negative number has the larger absolute value, the result is negative.

$8 - 5 = 3$ so $-8 + 5 = -3$

33. Subtract the smaller absolute value from the larger. Since the positive number has the larger absolute value, the result is positive.

$7 - 6 = 1$ so $7 + (-6) = 1$

35. Subtract the smaller absolute value from the larger. Since the negative number has the larger absolute value, the result is negative.

$42 - 20 = 22$ so $20 + (-42) = -22$

37. Subtract the smaller absolute value from the larger. Since the positive number has the larger absolute value, the result is positive.

$71 - 23 = 48$ so $71 + (-23) = 48$

39. Subtract the smaller absolute value from the larger. Since the positive number has the larger absolute value, the result is positive.

$479 - 122 = 357$ so $479 + (-122) = 357$

41. Subtract the smaller absolute value from the larger. Since the negative number has the larger absolute value, the result is negative.

$339 - 279 = 60$ so $-339 + 279 = -60$

43. $9 + (-3) + 5 + (-4)$
$= 6 + 5 + (-4)$
$= 11 + (-4)$
$= 7$

45. $6 + (-4) + (-13) + 7$
$= 2 + (-13) + 7$
$= -11 + 7$
$= -4$

47. $[-3 + (-4)] + (-5 + 2)$
$= [-7] + (-3)$
$= -10$

49. $(-1 + 34) + [16 + (-8)]$
$= 33 + 8$
$= 41$

51. Look for opposites (additive inverses):

$\underline{23} + (-5) + 3 + \underline{\underline{5}} + (-23) = 0 + 0 + 3 = 3$

53. Look for opposites (additive inverses):

$\underline{-10} + (-1) + \underline{10} + (-6) + \underline{\underline{1}} = 0 + 0 + (-6) = -\text{(}$

TRY IT YOURSELF

55. $-2 + 6 + (-1) = 4 + (-1) = 3$

57. $-7 + 0 = -7$

59. Subtract the smaller absolute value from the larger. Since the positive number has the larger absolute value, the result is positive.

$24 + (-15) = 9$

61. Add the absolute values. Since both numbers are negative, the result is negative.

$-435 + (-127) = -562$

63. Subtract the smaller absolute value from the larger. Since the positive number has the larger absolute value, the result is positive.

$-7 + 9 = 2$

65. $2 + (-2) = 0$

67. $2 + (-10 + 8) = 2 + (-2) = 0$

69. $\underline{-9} + \underline{\underline{1}} + (-2) + \underline{\underline{(-1)}} + \underline{9} = 0 + 0 + (-2) = -2$

71. $[6+(-4)]+[8+(-11)]$
$= 2+[-3]$
$= -1$

73. $(-4+8)+(-11+4)$
$= 4+(-7)$
$= -3$

75. $-675+(-456)+99$
$= -1,131+99$
$= -1,032$

77. $-6+(-7)+(-8)$
$= -13+(-8)$
$= -21$

79. $-2+[789+(-9,135)]$
$= -2+[-8,346]$
$= -8,348$

81. $-45+25=-20$

APPLICATIONS

83. Michigan: $-51°+163°=112°$

Minnesota: $-60°+174°=114°$

85. a. $-15,720 \, ft.$

b. $-15,720+3,220=-12,500 \, ft.$

87. a. $-9 \, ft.$

b. $-9+11=2 \, ft.$ above flood stage.

89. $-40+200+10+25$
$= 160+10+25$
$= 170+25$
$= 195°$

91. $-4+3+(-3)+3+4+2$
$= 0+0+3+2$
$= 5$

He has a 4% risk.

93. $-1,500+3,500+1,250$
$= 2,000+1,250$
$= 3,250 m.$

95. $5,889+927+(-2,928)+1,645+(-894)+715+(-6,321)$
$= 6,816+(-2,928)+1,645+(-894)+715+(-6,321)$
$= 3,888+1,645+(-894)+715+(-6,321)$
$= 5,533+(-894)+715+(-6,321)$
$= 4,639+715+(-6,321)$
$= 5,354+(-6,321)$
$= -967$

The entry should be ($967).

WRITING

97. No- if the negative number has a larger absolute value the sum will be negative.

99. Because we are beginning with a negative number and then taking away from it, so the result is still negative.

101. The integers are opposites, like 4 and -4.

REVIEW

103. a. $5+3+5+3=16 \, ft.$

b. $5 \cdot 3=15 \, ft^2$

105. $250=25 \cdot 10=5 \cdot 5 \cdot 5 \cdot 2$
$= 2 \cdot 5^3$

Section 2.3: Subtracting Integers

VOCABULARY

1. -8 is the opposite (or additive inverse) of 8.

3. To evaluate an expression means to find its value.

CONCEPTS

5. To subtract two integers, add the first integer to the opposite (additive inverse) of the integer to be subtracted.

7. Subtracting 3 is the same as adding -3.

9. We can find the <u>change</u> in a quantity by subtracting the earlier value from the later value.

11. a. The 3 is being subtracted.

    b. The -12 is being subtracted

13. $3-(-6)=3+6=9$

NOTATION

15. a. $-8-(-4)$

    b. $-4-(-8)$

17. $1-3-(-2)=1+(-3)+2$
    $=-2+2$
    $=0$

19. $(-8-2)-(-6)=[-8+(-2)]-(-6)$
    $=-10-(-6)$
    $=-10+6$
    $=-4$

GUIDED PRACTICE

21. $-4-3=-4+(-3)=-7$

23. $-5-5=-5+(-5)=-10$

25. $8-(-1)=8+1=9$

27. $11-(-7)=11+7=18$

29. $3-21=3+(-21)=-18$

31. $15-65=15+(-65)=-50$

33. a. $-11-(-1)=-11+1=-10$

    b. $-1-(-11)=-1+11=10$

35. a. $-16-(-41)=-16+41=25$

    b. $-41-(-16)=-41+16=-25$

Chapter 2 The Integers 33

37. $-4-(-4)-15=-4+4+(-15)$
    $=0+(-15)$
    $=-15$

39. $10-9-(-8)=10+(-9)+8$
    $=1+8$
    $=9$

41. $-1-(-3)-4=-1+3+(-4)$
    $=2+(-4)$
    $=-2$

43. $-5-8-(-3)=-5+(-8)+3$
    $=-13+3$
    $=-10$

45. $-1-(-4-6)=-1-[-4+(-6)]$
    $=-1-(-10)$
    $=-1+10$
    $=9$

47. $-42-(-16-14)=-42-[-16+(-14)]$
    $=-42-(-30)$
    $=-42+30$
    $=-12$

49. $-9-(6-7)=-9-[6+(-7)]$
    $=-9-(-1)$
    $=-9+1$
    $=-8$

51. $-8-(4-12)=-8-[4+(-12)]$
    $=-8-(-8)$
    $=-8+8$
    $=0$

53.
$$-(-5)+(-15)-6-(-48)$$
$$=5+(-15)+(-6)+48$$
$$=(5+48)+[-15+(-6)]$$
$$=53+(-21)$$
$$=32$$

55.
$$-(-3)+(-41)-7-(-19)$$
$$=3+(-41)+(-7)+19$$
$$=(3+19)+[-41+(-7)]$$
$$=22+(-48)$$
$$=-26$$

57. $-1,557-890=-2,447$

59. $-979-(-44,879)=-979+44,879=43,900$

TRY IT YOURSELF

61.
$$5-9-(-7)=5+(-9)+7$$
$$=-4+7$$
$$=3$$

63. $7-(-3)=7+3=10$

65. $-2-(-10)=-2+10=8$

67. $0-(-5)=0+5=5$

69.
$$(6-4)-(1-2)=2-(-1)$$
$$=2+1$$
$$=3$$

71. $-5-(-4)=-5+4=-1$

73.
$$-3-3-3=-3+(-3)+(-3)$$
$$=-6+(-3)$$
$$=-9$$

75.
$$-(-9)+(-20)-14-(-3)$$
$$=9+(-20)+(-14)+3$$
$$=(9+3)+[-20+(-14)]$$
$$=12+(-34)$$
$$=-22$$

77.
$$[-4+(-8)]-(-6)+15$$
$$=-12+6+15$$
$$=-6+15$$
$$=9$$

79. $-10-(-6)=-10+6=-4$

81. $-3-(-3)=-3+3=0$

83.
$$-8-[4-(-6)]$$
$$=-8-[4+6]$$
$$=-8-10$$
$$=-8+(-10)$$
$$=-18$$

85. $4-(-4)=4+4=8$

87.
$$(-6-5)-3+(-11)$$
$$=-6+(-5)+(-3)+(-11)$$
$$=-11+(-3)+(-11)$$
$$=-14+(-11)$$
$$=-25$$

APPLICATIONS

89. $-2,000-200=-2,000+(-200)=-2,200\ f$

91. $1,348-282=1,348+(-282)=1,066\ ft.$

93.
$$5-7-6=5+(-7)+(-6)$$
$$=-2+(-6)$$
$$=-8$$

Chapter 2 The Integers 35

95. $-1+(-6)+(-5)+8$
$=-7+(-5)+8$
$=-12+8$
$=-4\,yds.$

97. $15-25-30-40-60$
$=15+(-25)+(-30)+(-40)+(-60)$
$=-10+(-30)+(-40)+(-60)$
$=(-40)+(-40)+(-60)$
$=-80+(-60)$
$=-\$140$

99. Portland (142°), Barrow (135°), Kansas City (132°), Atlantic City (117°), Norfolk (107°)

101. $360-(-110)=360+110=470°F$

103. $-7-(-23)=-7+23=16$ point increase

WRITING

105. It means that we can think of all addition problems as subtraction problems and vice-versa.

107. By adding 4 to -11 and getting -7

REVIEW

109. a. 24,090

b. 6,000

111. $13\cdot12=156$ oranges needed

Section 2.4: Multiplying Integers

VOCABULARY

1. $\underset{\text{factor}}{-5}\cdot\underset{\text{factor}}{10}=\underset{\text{product}}{50}$

3. A positive integer and a negative integer are said to have different signs or unlike signs.

5. Associative property of multiplication: The way in which integers are grouped does not change their product.

CONCEPTS

7. Multiplication of integers is very much like multiplication of whole numbers. The only difference is that we must determine whether the answer is positive or negative.

9. To multiply a positive integer and a negative integer, multiply their absolute values. Then make the final answer negative.

11. The product of two integers with unlike/different signs is negative.

13. The product of any integer and 0 is 0.

15. a. $|-3|=3$

b. $|12|=12$

NOTATION

17. a. base: 8, exponent: 4

b. base -7, exponent: 9

19. $-3(-2)(-4)=6(-4)$
$=-24$

GUIDED PRACTICE

21. Multiply the absolute values. Since one number is negative, the result is negative.
$5\cdot3=15$ so $5(-3)=-15$

23. Multiply the absolute values. Since one number is negative, the result is negative.
$9\cdot2=18$ so $9(-2)=-18$

25. Multiply the absolute values. Since one number is negative, the result is negative.
$18\cdot4=72$ so $18(-4)=-72$

27. Multiply the absolute values. Since one number is negative, the result is negative.
$21\cdot6=126$ so $21(-6)=-126$

29. Multiply the absolute values. Since one number is negative, the result is negative.
$45 \cdot 37 = 1,665$ so $-45 \cdot 37 = -1,665$

31. Multiply the absolute values. Since one number is negative, the result is negative.
$94 \cdot 1,000 = 94,000$ so $-94 \cdot 1,000 = -94,00$

33. Multiply the absolute values. Since both numbers are negative, the result is positive.
$8 \cdot 7 = 56$ so $(-8)(-7) = 56$

35. Multiply the absolute values. Since both numbers are negative, the result is positive.
$7 \cdot 1 = 7$ so $-7(-1) = 7$

37. Multiply the absolute values. Since both numbers are negative, the result is positive.
$3 \cdot 52 = 156$ so $-3(-52) = 156$

39. Multiply the absolute values. Since both numbers are negative, the result is positive.
$6 \cdot 46 = 276$ so $-6(-46) = 276$

41. Multiply the absolute values. Since both numbers are negative, the result is positive.
$59 \cdot 33 = 1,947$ so $-59(-33) = 1,947$

43. Multiply the absolute values. Since both numbers are negative, the result is positive.
$60,000 \cdot 1,200 = 72,000,000$ so $-60,000 \cdot ($

45. Left to right: $6(-3)(-5) = -18(-5) = 90$

47. Left to right: $-5(10)(-3) = -50(-3) = 150$

49. First Pair, Last Pair:
$-2(-4)(6)(-8) = 8(-48) = -384$

51. First Pair, Last Pair:
$-8(-3)(7)(-2) = 24(-14) = -336$

53. Left to right: $-4(-2)(-6) = 8(-6) = -48$

55. Left to right: $-3(-9)(-3) = 27(-3) = -81$

57. First Pair, Last Pair:
$-1(-3)(-2)(-6) = 3(12) = 36$

59. First Pair, Last Pair:
$-9(-4)(-1)(-4) = 36(4) = 144$

61. $(-3)^3 = (-3)(-3)(-3) = 9(-3) = -27$

63. $(-2)^5 = (-2)(-2)(-2)(-2)(-2) = 4 \cdot 4(-2)$

65. $(-5)^4 = (-5)(-5)(-5)(-5) = 25(25) = 625$

67. $(-1)^8 = (-1)(-1)(-1) \cdots (-1) = 1 \cdot 1 \cdot 1 \cdot 1 = 1$

69. $(-7)^2 = (-7)(-7) = 49$
$-7^2 = -7 \cdot 7 = -49$

71. $(-12)^2 = (-12)(-12) = 144$
$-12^2 = -12 \cdot 12 = -144$

TRY IT YOURSELF

73. $6(-5)(2) = -30(2) = -60$

75. $-8(0) = 0$

77. $(-4)^3 = (-4)(-4)(-4) = 16(-4) = -64$

79. $(-2)10 = -20$

81. $-2(-3)(3)(-1) = 6(-3) = -18$

83. $-6(-10) = 60$

85. $-6(-4)(-2) = 24(-2) = -48$

87. $-42 \cdot 200,000 = -8,400,000$

89. $-5^4 = -5 \cdot 5 \cdot 5 \cdot 5 = -25 \cdot 25 = -625$

91. $-12(-12) = 144$

Chapter 2 The Integers   37

93. $(-1)^6 = (-1)(-1)(-1)(-1)(-1)(-1)$
$= 1 \cdot 1 \cdot 1 = 1$

95. $(-1)(-2)(-3)(-4)(-5) = 2(12)(-5)$
$= 24(-5) = -120$

APPLICATIONS

97. $8(-250) = -2000\, ft.$

99. a. high: 2, low: -3

b. high: $2 \cdot 2 = 4$, low: $2(-3) = -6$

101. a. $6(-67) = -402,000$ jobs

b. $9(-47) = -423,000$ jobs

c. $7(-83) = -581,000$ jobs

d. $6(-88) = -528,000$ jobs

103. $4(-81) = -324°F$

105. $6(-200) = -\$1,200$

107. $3(-6) = -18\, ft.$

109. $71,906(-3) = -\$215,718$

WRITING

111. Because we are really adding 5 negatives together (in this case) and the sum of any negative numbers is still negative.

113. Because the 1 doesn't have any effect on the value of the number.

REVIEW

115. 2, 3, 5, 7, 11, 13, 17, 19, 23, 29

117.
$$\begin{array}{r} 43 \\ 4\overline{)175} \\ -16 \\ \hline 15 \\ -12 \\ \hline 3 \end{array}$$

$175 \div 4 = 43 R3$

Section 2.5: Dividing Integers

VOCABULARY

1. $\underset{\text{dividend}}{12} \div \underset{\text{divisor}}{(-4)} = \underset{\text{quotient}}{-3}$

$\underset{\text{divisor}}{\dfrac{\overset{\text{dividend}}{12}}{-4}} = \underset{\text{quotient}}{-3}$

3. $\dfrac{-3}{0}$ is division <u>by</u> 0 and $\dfrac{0}{-3}$ is division <u>of</u> 0.

CONCEPTS

5. a. $-5(5) = -25$

b. $6(-6) = -36$

c. $0(-15) = 0$

7. a. The quotient of two integers that have the same *(like)* signs is <u>positive</u>.

b. The quotient of two integers that have different *(unlike)* signs is <u>negative</u>.

9. a. If 0 is divided by ay nonzero integer, the quotient is <u>0</u>.

b. Division of any nonzero number by 0 is <u>undefined</u>.

11. a. always true

    b. sometimes true

    c. always true

GUIDED PRACTICE

13. Divide the absolute values: the signs are opposite so the result is negative.

    $\dfrac{14}{2} = 7$ so $\dfrac{-14}{2} = -7$

    $-7 \cdot 2 = -14$ ☺

15. Divide the absolute values: the signs are opposite so the result is negative.

    $\dfrac{20}{5} = 4$ so $\dfrac{-20}{5} = -4$

    $-4 \cdot 5 = -20$ ☺

17. Divide the absolute values: the signs are opposite so the result is negative.

    $36 \div 6 = 6$ so $36 \div (-6) = -6$

    $-6 \cdot (-6) = 36$ ☺

19. Divide the absolute values: the signs are opposite so the result is negative.

    $24 \div 3 = 8$ so $24 \div (-3) = -8$

    $-8(-3) = 24$ ☺

21. Divide the absolute values: the signs are opposite so the result is negative.

    $\dfrac{264}{12} = 22$ so $\dfrac{264}{-12} = -22$

    $-22 \cdot (-12) = 264$ ☺

23. Divide the absolute values: the signs are opposite so the result is negative.

    $\dfrac{702}{18} = 39$ so $\dfrac{702}{-18} = -39$

    $-39 \cdot (-18) = 702$ ☺

25. Divide the absolute values: the signs are opposite so the result is negative.

    $9{,}000 \div 300 = 30$ so $-9{,}000 \div 300 = -30$

    $-30 \cdot 300 = -9{,}000$ ☺

27. Divide the absolute values: the signs are opposite so the result is negative.

    $250{,}000 \div 5{,}000 = 50$ so $-250{,}000 \div 5{,}000$

    $-50 \cdot 5{,}000 = -250{,}000$ ☺

29. Divide the absolute values: the signs are the same so the result is positive.

    $\dfrac{8}{4} = 2$ so $\dfrac{-8}{-4} = 2$

    $2(-4) = -8$ ☺

31. Divide the absolute values: the signs are the same so the result is positive.

    $\dfrac{45}{9} = 5$ so $\dfrac{-45}{-9} = 5$

    $5(-9) = -45$ ☺

33. Divide the absolute values: the signs are the same so the result is positive.

    $63 \div 7 = 9$ so $-63 \div (-7) = 9$

    $9(-7) = -63$ ☺

35. Divide the absolute values: the signs are the same so the result is positive.

    $32 \div 8 = 4$ so $-32 \div (-8) = 4$

    $4(-8) = -32$ ☺

37. Divide the absolute values: the signs are the same so the result is positive.

    $\dfrac{400}{25} = 16$ so $\dfrac{-400}{-25} = 16$

$16(-25) = -400$ ☺

39. Divide the absolute values: the signs are the same so the result is positive.

$\dfrac{651}{31} = 21$ so $\dfrac{-651}{-31} = 21$

$21(-31) = -651$ ☺

41. Divide the absolute values: the signs are the same so the result is positive.

$800 \div 20 = 40$ so $-800 \div (-20) = 40$

$40(-20) = -800$ ☺

43. Divide the absolute values: the signs are the same so the result is positive.

$15{,}000 \div 30 = 500$ so $-15{,}000 \div (-30) = 50$

$500(-30) = -15{,}000$ ☺

45. a. undefined

b. 0

47. a. 0

b. undefined

TRY IT YOURSELF

49. Divide the absolute values: the signs are the same so the result is positive.

$36 \div 12 = 3$ so $-36 \div (-12) = 3$

51. Divide the absolute values: the signs are opposite so the result is negative.

$\dfrac{425}{25} = 17$ so $\dfrac{425}{-25} = -17$

53. $0 \div (-16) = 0$

Chapter 2 The Integers 39

55. Divide the absolute values: the signs are opposite so the result is negative.

$45 \div 9 = 5$ so $-45 \div 9 = -5$

57. Divide the absolute values: the signs are opposite so the result is negative.

$2{,}500 \div 500 = 5$ so $-2{,}500 \div (500) = -5$

59. $\dfrac{-6}{0}$ is undefined.

61. Divide the absolute values: the signs are opposite so the result is negative.

$\dfrac{19}{1} = 19$ so $\dfrac{19}{-1} = -19$

63. Divide the absolute values: the signs are the same so the result is positive.

$23 \div 23 = 1$ so $-23 \div (-23) = 1$

65. Divide the absolute values: the signs are opposite so the result is negative.

$\dfrac{40}{2} = 20$ so $\dfrac{40}{-2} = -20$

67. Divide the absolute values: the signs are opposite so the result is negative.

$9 \div 9 = 1$ so $9 \div (-9) = -1$

69. Divide the absolute values: the signs are the same so the result is positive.

$\dfrac{10}{1} = 10$ so $\dfrac{-10}{-1} = 10$

71. Divide the absolute values: the signs are opposite so the result is negative.

$\dfrac{888}{37} = 24$ so $\dfrac{-888}{37} = -24$

73. Divide the absolute values: the signs are opposite so the result is negative.

$$\frac{3,000}{100} = 30 \text{ so } \frac{3,000}{-100} = -30$$

75. Divide the absolute values: the signs are opposite so the result is negative.

$$8 \div 2 = 4 \text{ so } 8 \div (-2) = -4$$

77. $\dfrac{-13,550}{25} = -542$

79. $\dfrac{27,778}{-17} = -1,634$

APPLICATIONS

81. $-210 \div 6 = -\$35 / wk.$

83. $-3,030 \div 3 = -1,010 \, ft.$

85. $30 - 72 = -42°$
    $-42 \div 6 = -7° / min$

87. $-12 \div 2 = -6$ (6 games behind)

89. $-300 \div 20 = -\$15$ per pair

91. $-4,335 \div 255 = -\$17$ per share

WRITING

93. The quotient of two negatives is positive because we are really dividing top and bottom by negative 1.

95. To check that $20 \div 4 = 5$, verify $4 \cdot 5 = 20$.

97. The quotient of two integers that have the same signs is positive and the quotient of two integers that have different signs is negative.

REVIEW

99.
$$5^2 \left(\frac{2 \cdot 3^2}{6} \right)^2 - 7(2)$$
$$= 25 \left(\frac{2 \cdot 9}{6} \right)^2 - 14$$
$$= 25 \left(\frac{18}{6} \right)^2 - 14$$
$$= 25(3)^2 - 14$$
$$= 25 \cdot 9 - 14$$
$$= 225 - 14$$
$$= 211$$

101. associative property of addition

103. no

Section 2.6: Order of Operations and Estimation

VOCABULARY

1. To evaluate expressions that contain more than one operation, we use the order of operations rule.

3. In the expression $-9 - 2\left[-5 - 6(-3-1)\right]$, the parentheses are the inner most grouping symbols and the brackets are the outer most grouping symbols.

CONCEPTS

5. a. square, multiply, subtract

   b. multiply, cube, subtract, add

   c. subtract, multiply, add

   d. square, multiply

NOTATION

7. parentheses, brackets, absolute values, fraction bar

9.
$$-8 - 5(-2)^2 = -8 - 5(4)$$
$$= -8 - 20$$
$$= -8 + (-20)$$
$$= -28$$

11. $-9+5[-4\cdot 2+7] = -9+5[-8+7]$
$= -9+5[-1]$
$= -9+(-5)$
$= -14$

GUIDED PRACTICE

13. $-2(-3)^2-(-8)$
$= -2(9)-(-8)$
$= -18+8$
$= -10$

15. $-5(-4)^2-(-18)$
$= -5(16)-(-18)$
$= -80+18$
$= -62$

17. $9(7)+(-6)(-2)(-4)$
$= 63+12(-4)$
$= 63-48$
$= 15$

19. $8(6)+(-2)(-9)(-2)$
$= 48+18(-2)$
$= 48-36$
$= 12$

21. $30\div(-5)2$
$= -6\cdot 2$
$= -12$

23. $60\div(-3)4$
$= -20\cdot 4$
$= -80$

25. $-6^2-(-6)^2$
$= -36-36$
$= -72$

27. $-10^2-(-10)^2$
$= -100-100$
$= -200$

29. $-14+2(-9+6\cdot 3)$
$= -14+2(-9+18)$
$= -14+2\cdot 9$
$= -14+18$
$= 4$

31. $-23+3(-15+8\cdot 4)$
$= -23+3(-15+32)$
$= -23+3(17)$
$= -23+51$
$= 28$

33. $77-2\left[-6+(3-9)^2\right]$
$= 77-2\left[-6+(-6)^2\right]$
$= 77-2[-6+36]$
$= 77-2\cdot 30$
$= 77-60$
$= 17$

35. $99-4\left[-9+(6-10)^2\right]$
$= 99-4\left[-9+(-4)^2\right]$
$= 99-4[-9+16]$
$= 99-4\cdot 7$
$= 99-28$
$= 71$

37. $-\left[4-\left(3^3+\dfrac{22}{-11}\right)\right]$
$= -[4-(27+(-2))]$
$= -[4-25]$
$= -(-21)$
$= 21$

39. $-\left[50-\left(5^3+\dfrac{50}{-2}\right)\right]$
$=-\left[50-(125+(-25))\right]$
$=-[50-100]$
$=-(-50)$
$=50$

41. $\dfrac{-24+3(-4)}{42-(-6)^2}$
$=\dfrac{-24-12}{42-36}$
$=\dfrac{-36}{6}$
$=-6$

43. $\dfrac{-38+11(-2)}{69-(-8)^2}$
$=\dfrac{-38-22}{69-64}$
$=\dfrac{-60}{5}$
$=-12$

45. a. $|-6(2)|=|-12|=12$

   b. $|-12+7|=|-5|=5$

47. a. $|15(-4)|=|-60|=60$

   b. $|16+(-30)|=|-14|=14$

49. $16-6|-2-1|$
$=16-6|-3|$
$=16-6\cdot 3$
$=16-18$
$=-2$

51. $17-2|-6-4|$
$=17-2|-10|$
$=17-2\cdot 10$
$=17-20$
$=-3$

53. $-379+(-13)+287+(-671)$
$\approx -380+(-10)+290+(-670)$
$=-390+290+(-670)$
$=-100+(-670)$
$=-770$

55. $-3{,}887+(-5{,}806)+4{,}701$
$\approx -3{,}900+(-5{,}800)+4{,}700$
$=-9{,}700+4{,}700$
$=-5{,}000$

TRY IT YOURSELF

57. $(-3)^2-4^2$
$=9-16$
$=-7$

59. $3^2-4(-2)(-1)$
$=9+8(-1)$
$=9-8$
$=1$

61. $|-3\cdot 4+(-5)|$
$=|-12+(-5)|$
$=|-17|$
$=17$

63. $(2-5)(5+2)$
$=(-3)(7)$
$=-21$

65. $6 + \dfrac{25}{-5} + 6 \cdot 3$
$= 6 - 5 + 18$
$= 1 + 18$
$= 19$

67. $\dfrac{-6 - 2^3}{-2 - (-4)}$
$= \dfrac{-6 - 8}{-2 + 4}$
$= \dfrac{-14}{2}$
$= -7$

69. $-12 \div (-2) 2$
$= 6 \cdot 2$
$= 12$

71. $-16 - 4 \div (-2)$
$= -16 + 2$
$= -14$

73. $-\left| 2 \cdot 7 - (-5)^2 \right|$
$= -|14 - 25|$
$= -|-11|$
$= -11$

75. $|-4 - (-6)|$
$= |-4 + 6|$
$= |2|$
$= 2$

77. $(7-5)^2 - (1-4)^2$
$= 2^2 - (-3)^2$
$= 4 - 9$
$= -5$

79. $-1(2^2 - 2 + 1^2)$
$= -1(4 - 2 + 1)$
$= -1(2 + 1)$
$= -1(3)$
$= -3$

81. $\dfrac{-5 - 5}{1^4 + 1^5}$
$= \dfrac{-10}{1 + 1}$
$= \dfrac{-10}{2}$
$= -5$

83. $-50 - 2(-3)^3 (4)$
$= -50 - 2(-27)(4)$
$= -50 + 54 \cdot 4$
$= -50 + 216$
$= 166$

85. $-6^2 + 6^2 = -36 + 36 = 0$

87. $3\left(\dfrac{-18}{3} \right) - 2(-2)$
$= 3(-6) + 4$
$= -18 + 4$
$= -14$

89. $2|1 - 8| \cdot |-8|$
$= 2|-7| \cdot 8$
$= 2 \cdot 7 \cdot 8$
$= 112$

91.
$$\frac{2+3[5-(1-10)]}{|2(-8+2)+10|}$$
$$=\frac{2+3[5-(-9)]}{|2(-6)+10|}$$
$$=\frac{2+3[5+9]}{|-12+10|}$$
$$=\frac{2+3(14)}{|-2|}$$
$$=\frac{2+42}{2}$$
$$=\frac{44}{2}$$
$$=22$$

93. $-2+|6-4^2|$
$=-2+|6-16|$
$=-2+|-10|$
$=-2+10$
$=8$

95. $\dfrac{-4(-5)-2}{3-3^2}$
$=\dfrac{20-2}{3-9}$
$=\dfrac{18}{-6}$
$=-3$

APPLICATIONS

97. $-70+60-230-30-130=-400$ points

99. $12(3)+3(-4)+5(-1)$
$=36-12-5=19$

101. $\dfrac{-12+(-15)+(-5)+(-16)}{6}$
$=\dfrac{-48}{6}$
$=-\$8 \, million$

103. Last five:
$$\frac{-164+(-107)+(-22)+70+123}{5}=\$-20$$

Last four: $\dfrac{(-107)+(-22)+70+123}{4}=\$16$

It would be better to refer to the last 4 years because there is a surplus.

105. a. $-30-60=-90 \, ft.$

b. $10(60)=\$600$ lost

c. $-8,000 \div 20 = -400 \, ft.$

WRITING

107. It is necessary so that everyone gets the same answer from the same problem.

109. a. The division should have been done before the multiplication.

b. The multiplication of the 8 and the absolute value should happen before adding the -1.

REVIEW

111. a. $4+(-7)=-3$

b. $2-6=-4$

113. $7(140)=980 \, lbs.$ - it is fine for them to be on the elevator.

Chapter 2 Review

1. $\{...,-3,-2,-1,0,1,2,3,...\}$

3. $-33 \, ft$

5. a. $0>7$ b. $-20<-19$

7. a. $|5|=5$ b. $|-43|=43$ c. $|0|=0$

9. a. $-|12|=-12$

   b. $-(-12)=12$

   c. $-0=0$

11.
Position	Player	Score to Par
1	Helen Alfredsson	-12
2	Yani Tseng	-9
3	Laura Diaz	-8
4	Karen Stupples	-7
5	Young Kim	-6
6	Shanshan Feng	-5

13. Add the absolute values. Since both numbers are negative, the result is negative.

    $6+4=10$ so $-6+(-4)=-10$

15. Subtract the smaller absolute value from the larger. Since the positive number has the larger absolute value, the result is positive.

    $60-28=32$ so $-28+60=32$

17. $-8+8=0$

19. $-1+(-4)+(-3)=-5+(-3)=-8$

21. $[7+(-9)]+(-4+16)$
    $=-2+12=10$

23. $-4+0=-4$

25. $-2+(-1)+(-76)+1+2$
    $=-2+2+(-1)+1+(-76)$
    $=0+0+(-76)$
    $=-76$

27. $-102+73+(-345)$
    $=-29+(-345)$
    $=-374$

29. a. 11   b. -4

31. a. $-100\,ft$

    b. $-100+16+18=-34\,ft$

33. Subtracting an integer is the same as adding the opposite of that integer.

35. $5-8=5+(-8)=-3$

37. $-4-(-8)=-4+8=4$

39. $-6-106=-6+(-106)=-112$

41. $0-37=0+(-37)=-37$

43. $12-2-(-6)$
    $=12+(-2)+6$
    $=10+6$
    $=16$

45. $-9-7+12$
    $=-9+(-7)+12$
    $=-16+12$
    $=-4$

47. $1-(2-7)$
    $=1-(-5)$
    $=1+5$
    $=6$

49. $-70-[(-6)-2]$
    $=-70-(-8)$
    $=-70+8$
    $=-62$

51. $-(-5)+(-28)-2-(-100)$
    $=5+(-28)+(-2)+100$
    $=-23+(-2)+100$
    $=-25+100$
    $=75$

53. $-150-75=-150+(-75)=-225\,ft$

55. $28-(-16)=28+16=44$ points

57. Multiply the absolute values. Since one number is negative, the result is negative.

$7 \cdot 2 = 14$ so $7(-2) = -14$

59. Multiply the absolute values. Since both numbers are negative, the result is positive.

$23 \cdot 14 = 322$ so $(-23)(-14) = 322$

61. Multiply the absolute values. Since one number is negative, the result is negative.

$1 \cdot 25 = 25$ so $-1 \cdot 25 = -25$

63. Multiply the absolute values. Since one number is negative, the result is negative.

$4,000 \cdot 17,000 = 68,000,000$
so $-4,000(17,000) = -68,000,000$

65. $(-6)(-2)(-3) = (12)(-3) = -36$

67. $(-3)(4)(2)(-5) = -12(2)(-5)$
$= (-24)(-5) = 120$

69. $-15(-30) = 450$

71. State Treasurer: $2(-130) = -\$260$

Governor: $3(-130) = -\$390$

73. $(-5)^3 = (-5)(-5)(-5)$
$= 25(-5) = -125$

75. $(-8)^4 = (-8)(-8)(-8)(-8)$
$= 64(64) = 4,096$

77. Since the exponent is odd, the result will be negative.

79. We know that $\dfrac{-15}{5} = -3$ because
$-3(5) = -15$.

81. Divide the absolute values: the signs are opposite so the result is negative.

$\dfrac{25}{5} = 5$ so $\dfrac{25}{-5} = -5$

83. Divide the absolute values: the signs are the same so the result is positive.

$64 \div 8 = 8$ so $(-64) \div (-8) = 8$

85. Divide the absolute values: the signs are the same so the result is positive.

$\dfrac{10}{1} = 10$ so $\dfrac{-10}{-1} = 10$

87. Divide the absolute values: the signs are opposite so the result is negative.

$150,000 \div 3,000 = 50$
so $-150,000 \div 3,000 = -50$

89. Divide the absolute values: the signs are the same so the result is positive.

$\dfrac{1,058}{46} = 23$ so $\dfrac{-1,058}{-46} = 23$

91. $\dfrac{0}{-5} = 0$

93. $-96 \div 3 = -32$

95. $-12 \div 6 = -2$ min

97. $2 + 4(-6)$
$= 2 + (-24)$
$= -22$

99. $65 - 8(9) - (-47)$
$= 65 - 72 + 47$
$= -7 + 47$
$= 40$

101.
$$-2(5)(-4)+\frac{|-9|}{3^2}$$
$$=-10(-4)+\frac{9}{9}$$
$$=40+1$$
$$=41$$

103.
$$-12-(8-9)^2$$
$$=-12-(-1)^2$$
$$=-12-1$$
$$=-13$$

105.
$$-4\left(\frac{15}{-3}\right)-2^3$$
$$=-4(-5)-8$$
$$=20-8$$
$$=12$$

107.
$$-20+2\left[12-(-7+5)^2\right]$$
$$=-20+2\left[12-(-2)^2\right]$$
$$=-20+2[12-4]$$
$$=-20+2[8]$$
$$=-20+16$$
$$=-4$$

109.
$$\frac{2\cdot 5+(-6)}{-3-1^5}$$
$$=\frac{10+(-6)}{-3-1}$$
$$=\frac{4}{-4}$$
$$=-1$$

111.
$$-\left[1-\left(2^3+\frac{100}{-50}\right)\right]$$
$$=-[1-(8+(-2))]$$
$$=-[1-6]$$
$$=-[-5]$$
$$=5$$

113.
$$-4,500+7,900+2,100+(-3,200)$$
$$=3,400+2,100+(-3,200)$$
$$=5,500+(-3,200)$$
$$=2,300$$

Chapter 2 Test

1. a. $\{\ldots,-5,-4,-3,-2,-1,0,1,2,3,4,5,\ldots\}$ is called the set of <u>integers</u>.

   b. The symbols > and < are called <u>inequality</u> symbols.

   c. The <u>absolute value</u> of a number is the distance between the number and 0 on the number line.

   d. Two numbers that are the same distance from 0 on the number line but on opposite sides of it, are called <u>opposites</u>.

   e. In the expression $(-3)^5$, the <u>base</u> is -3 and 5 is the <u>exponent</u>.

3. a. true b. true

   c. false d. false

   e. true

5.

7. a. $-7-6=-7+(-6)=-13$
   b. $-7-(-6)=-7+6=-1$
   c. $82-(-109)=82+109=191$
   d. $0-15=0+(-15)=-15$
   e. $-60-50-40=-110-40=-150$

9. $\dfrac{-20}{-4}=5 \Leftrightarrow 5(-4)=-20$

11. a. $-27+15=-12$
    b. $-1-(-19)=-1+19=18$
    c. $\dfrac{-28}{-7}=\dfrac{28}{7}=4$
    d. $10(-8)=-80$

13. a. $\dfrac{-21}{0}$ is undefined
    b. $\dfrac{-5}{1}=-5$
    c. $\dfrac{0}{-6}=0$
    d. $\dfrac{-18}{-18}=1$

15. $4-(-3)^2-(-6)$
    $=4-9+6$
    $=-5+6$
    $=1$

17. $-3+\left(\dfrac{-16}{4}\right)-3^3$
    $=-3+(-4)-27$
    $=-7-27$
    $=-34$

19. $\dfrac{4(-6)-4^2+(-2)}{-3-4\cdot 1^5}$
    $=\dfrac{-24-16+(-2)}{-3-4\cdot 1}$
    $=\dfrac{-40+(-2)}{-3-4}$
    $=\dfrac{-42}{-7}$
    $=6$

21. $21-9|-3-4+2|$
    $=21-9|-5|$
    $=21-9\cdot 5$
    $=21-45$
    $=-24$

23. $-6\cdot 12=-72°F$

25. $436-282=154\,ft$ difference in elevations.

27. In this hand he has $3+10+9+1=23$ points.

    His new score is $8-23=-15$ points.

Chapters 1 – 2 Cumulative Review

1. a. 7 millions
   b. 3
   c. 7,326,500
   d. 7,330,000

3.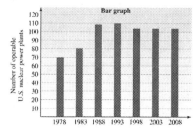

5. $\overset{1\ 1}{1,237}$
    549
    + 68
    ———
    1,854

7. $\overset{5\ \ 17\ \ 6\ \ 17}{\cancel{6},\cancel{3}\cancel{7}\cancel{5}}$
    −2,569
    ———
    3,806

9. $\overset{8\ \ 14\ \ 9\ \ 16}{\cancel{3}\cancel{9},\cancel{5}\cancel{0}\cancel{6}}$
    −1,729
    ———
    37,777

11. $132 + 237 = 369$ - it is incorrect.

13.
    435
    $\times 27$
    ———
    3,045
    +8,700
    ———
    11,745

15. $31 \cdot 7 = 217$

    attaching 5 zeros gives $21,700,000$

17. $P = 17 + 35 + 17 + 35$
    $= 52 + 52 = 104\,ft$
    $A = 17 \cdot 35 = 595\,ft^2$

19. $8\overline{)701}$ quotient 87
    −64
    ———
    61
    −56
    ———
    5
    $87\,R\,5$

21. $38\overline{)17746}$ quotient 467
    −152
    ———
    257
    −228
    ———
    296
    −296
    ———
    0

23. $218 \cdot 9 = 1,962$ - it is correct!

25. $15 \cdot 16 = 240\,oz$
    $240 \div 4 = 60$

    He can make 60 dinner rolls.

27. a. prime, odd

    b. composite, even

    c. neither, even

    d. neither, odd

29. $11 \cdot 11 \cdot 11 \cdot 11 = 11^4$

31. $8 = 2 \cdot 2 \cdot 2$
    $12 = 2 \cdot 2 \cdot 3$
    $LCM(8,12) = 2 \cdot 2 \cdot 2 \cdot 3 = 24$

33. $30 = 2 \cdot 3 \cdot 5$
    $48 = 2 \cdot 2 \cdot 2 \cdot 2 \cdot 3$
    $GCF(30,48) = 2 \cdot 3 = 6$

35. $16 + 2\left[14 - 3(5-4)^2\right]$
    $= 16 + 2\left[14 - 3(1)^2\right]$
    $= 16 + 2[14 - 3]$
    $= 16 + 2[11]$
    $= 16 + 22$
    $= 38$

37. $$\frac{4^2 - 2 \cdot 3}{2 + (3^2 - 3 \cdot 2)}$$
$$= \frac{16 - 6}{2 + 9 - 6}$$
$$= \frac{10}{5}$$
$$= 2$$

39. a.

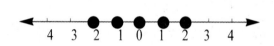

b.

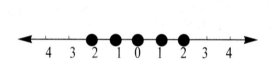

41. $3 - (-18) = 3 + 18 = 21$

43. $7(-39) = -273°C$

45. $5 + (-3)(-7)(-2)$
$= 5 + 21(-2)$
$= 5 + (-42)$
$= -37$

47. $$\frac{10 - (-5)}{1 - 2 \cdot 3}$$
$$= \frac{10 + 5}{1 - 6}$$
$$= \frac{15}{-5}$$
$$= -3$$

49. $3^4 + 6(-12 + 5 \cdot 4)$
$= 81 + 6(-12 + 20)$
$= 81 + 6(8)$
$= 81 + 48$
$= 129$

51. $2\left(\frac{-12}{3}\right) + 3(-5)$
$= 2(-4) + 3(-5)$
$= -8 + (-15)$
$= -23$

53. $-\left|\frac{45}{-9} - (-9)\right|$
$= -|-5 + 9|$
$= -|4|$
$= -4$

55. $\dfrac{1{,}150}{12} \approx 95.83 \, ft$

CHAPTER 3 Fractions and Mixed Numbers

Section 3.1: An Introduction to Fractions

VOCABULARY

1. A <u>fraction</u> describes the number of equal parts of a whole.

3. If the numerator of a fraction is less than its denominator, the fraction is called a <u>proper</u> fraction. If the numerator of a fraction is greater than or equal to its denominator it is called an <u>improper</u> fraction.

5. Two fractions are <u>equivalent</u> if they represent the same number.

7. Writing a fraction as an equivalent fraction with a larger denominator is called <u>building</u> the fraction.

CONCEPTS

9. Equivalent fractions: $\dfrac{2}{6} = \dfrac{1}{3}$

11. a. $\dfrac{37}{24}$ improper fraction   b. $\dfrac{1}{3}$ proper fraction

    c. $\dfrac{71}{100}$ proper fraction   d. $\dfrac{9}{9}$ improper

13. 5

15. Multiplying fractions: To multiply two fractions, multiply the <u>numerators</u> and multiply the denominators.

NOTATION

17. $\dfrac{-7}{8}, -\dfrac{7}{8}$

19. $\dfrac{1}{6} = \dfrac{1}{6} \cdot \dfrac{3}{3} = \dfrac{1 \cdot 3}{6 \cdot 3} = \dfrac{3}{18}$

GUIDED PRACTICE

21. numerator 4; denominator 5

23. numerator 17; denominator 10

25. $\dfrac{3}{4}$ shaded, $\dfrac{1}{4}$ not shaded

27. $\dfrac{5}{8}$ shaded, $\dfrac{3}{8}$ not shaded

29. $\dfrac{2}{5}$ shaded, $\dfrac{3}{5}$ not shaded

31. $\dfrac{7}{12}$ shaded, $\dfrac{5}{12}$ not shaded

33. a. $\dfrac{4}{1} = 4$ b. $\dfrac{8}{8} = 1$

    c. $\dfrac{0}{12} = 0$ d. $\dfrac{1}{0}$ undefined

35. a. $\dfrac{5}{0}$ undefined   b. $\dfrac{0}{50} = 0$

    c. $\dfrac{33}{33} = 1$ d. $\dfrac{75}{1} = 75$

37. $\dfrac{7}{8} = \dfrac{7}{8} \cdot \dfrac{5}{5} = \dfrac{7 \cdot 5}{8 \cdot 5} = \dfrac{35}{40}$

39. $\dfrac{4}{9} = \dfrac{4}{9} \cdot \dfrac{3}{3} = \dfrac{4 \cdot 3}{9 \cdot 3} = \dfrac{12}{27}$

41. $\dfrac{5}{6} = \dfrac{5}{6} \cdot \dfrac{9}{9} = \dfrac{5 \cdot 9}{6 \cdot 9} = \dfrac{45}{54}$

43. $\dfrac{2}{7} = \dfrac{2}{7} \cdot \dfrac{2}{2} = \dfrac{2 \cdot 2}{7 \cdot 2} = \dfrac{4}{14}$

45. $\dfrac{1}{2} = \dfrac{1}{2} \cdot \dfrac{15}{15} = \dfrac{1 \cdot 15}{2 \cdot 15} = \dfrac{15}{30}$

47. $\dfrac{11}{16} = \dfrac{11}{16} \cdot \dfrac{2}{2} = \dfrac{11 \cdot 2}{16 \cdot 2} = \dfrac{22}{32}$

49. $\dfrac{5}{4} = \dfrac{5}{4} \cdot \dfrac{7}{7} = \dfrac{5 \cdot 7}{4 \cdot 7} = \dfrac{35}{28}$

51. $\dfrac{16}{15} = \dfrac{16}{15} \cdot \dfrac{3}{3} = \dfrac{16 \cdot 3}{15 \cdot 3} = \dfrac{48}{45}$

53. $4 = \dfrac{4}{1} \cdot \dfrac{9}{9} = \dfrac{4 \cdot 9}{1 \cdot 9} = \dfrac{36}{9}$

55. $6 = \dfrac{6}{1} \cdot \dfrac{8}{8} = \dfrac{6 \cdot 8}{1 \cdot 8} = \dfrac{48}{8}$

57. $3 = \dfrac{3}{1} \cdot \dfrac{5}{5} = \dfrac{3 \cdot 5}{1 \cdot 5} = \dfrac{15}{5}$

59. $14 = \dfrac{14}{1} \cdot \dfrac{2}{2} = \dfrac{14 \cdot 2}{1 \cdot 2} = \dfrac{28}{2}$

61. a. No   b. Yes

63. a. Yes   b. No

65. $\dfrac{6}{9} = \dfrac{2 \cdot 3}{3 \cdot 3} = \dfrac{2 \cdot \cancel{3}}{3 \cdot \cancel{3}} = \dfrac{2}{3}$

67. $\dfrac{16}{20} = \dfrac{2 \cdot 2 \cdot 2 \cdot 2}{2 \cdot 2 \cdot 5} = \dfrac{\cancel{2} \cdot \cancel{2} \cdot 2 \cdot 2}{\cancel{2} \cdot \cancel{2} \cdot 5} = \dfrac{4}{5}$

69. $\dfrac{7}{21} = \dfrac{7}{3 \cdot 7} = \dfrac{\cancel{7}}{3 \cdot \cancel{7}} = \dfrac{1}{3}$

71. $\dfrac{2}{48} = \dfrac{2}{2 \cdot 2 \cdot 2 \cdot 2 \cdot 3} = \dfrac{\cancel{2}}{\cancel{2} \cdot 2 \cdot 2 \cdot 2 \cdot 3} = \dfrac{1}{24}$

73. $\dfrac{16}{17}$ in lowest terms

75. $\dfrac{36}{96} = \dfrac{2 \cdot 2 \cdot 3 \cdot 3}{2 \cdot 2 \cdot 2 \cdot 2 \cdot 2 \cdot 3} = \dfrac{\cancel{2} \cdot \cancel{2} \cdot 3 \cdot \cancel{3}}{\cancel{2} \cdot \cancel{2} \cdot 2 \cdot 2 \cdot 2 \cdot \cancel{3}} = \dfrac{3}{8}$

77. $\dfrac{55}{62}$ in simplest form

79. $\dfrac{50}{55} = \dfrac{2 \cdot 5 \cdot 5}{5 \cdot 11} = \dfrac{2 \cdot \cancel{5} \cdot 5}{\cancel{5} \cdot 11} = \dfrac{10}{11}$

81. $\dfrac{60}{108} = \dfrac{2 \cdot 2 \cdot 3 \cdot 5}{2 \cdot 2 \cdot 3 \cdot 3 \cdot 3} = \dfrac{\cancel{2} \cdot \cancel{2} \cdot \cancel{3} \cdot 5}{\cancel{2} \cdot \cancel{2} \cdot \cancel{3} \cdot 3 \cdot 3} = \dfrac{5}{9}$

83. $\dfrac{180}{210} = \dfrac{6 \cdot \cancel{30}}{7 \cdot \cancel{30}} = \dfrac{6}{7}$

85. $\dfrac{234}{306} = \dfrac{13 \cdot \cancel{18}}{17 \cdot \cancel{18}} = \dfrac{13}{17}$

87. $\dfrac{15}{6} = \dfrac{\cancel{3} \cdot 5}{2 \cdot \cancel{3}} = \dfrac{5}{2}$

89. $\dfrac{420}{144} = \dfrac{\cancel{12} \cdot 35}{\cancel{12} \cdot 12} = \dfrac{35}{12}$

91. $-\dfrac{4}{68} = -\dfrac{\cancel{4}}{\cancel{4} \cdot 17} = -\dfrac{1}{17}$

93. $-\dfrac{90}{105} = -\dfrac{6 \cdot \cancel{15}}{7 \cdot \cancel{15}} = -\dfrac{6}{7}$

95. $-\dfrac{16}{26} = -\dfrac{\cancel{2} \cdot 8}{\cancel{2} \cdot 13} = -\dfrac{8}{13}$

TRY IT YOURSELF

97. $\dfrac{2}{14} = \dfrac{1}{7}$; $\dfrac{6}{36} = \dfrac{1}{6}$ not equivalent

99. $\dfrac{22}{34} = \dfrac{11}{17}$; $\dfrac{33}{51} = \dfrac{11}{17}$ equivalent

APPLICATIONS

101. a. 32

b. $\dfrac{5}{32}$

103. a. 16

b. $\dfrac{10}{16} = \dfrac{\cancel{2} \cdot 5}{\cancel{2} \cdot 8} = \dfrac{5}{8}$

105. a. 28 Democrat, 22 Republican

b. $\dfrac{28}{50} = \dfrac{\cancel{2} \cdot 14}{\cancel{2} \cdot 25} = \dfrac{14}{25}$

c. $\dfrac{22}{50} = \dfrac{\cancel{2} \cdot 11}{\cancel{2} \cdot 25} = \dfrac{11}{25}$

107. a. 20

b. $\dfrac{8}{20} = \dfrac{2 \cdot \cancel{4}}{\cancel{4} \cdot 5} = \dfrac{2}{5}$ sold

$\dfrac{12}{20} = \dfrac{3 \cdot \cancel{4}}{\cancel{4} \cdot 5} = \dfrac{3}{5}$ not sold

109.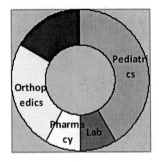

WRITING

111. Equivalent fractions are those that reduce to the same value. For example, $\dfrac{3}{6}$ and $\dfrac{4}{8}$ are equivalent since both reduce to $\dfrac{1}{2}$.

113. Because the pieces aren't divided up evenly.

115. a. Building a fraction: The numerator and denominator must be multiplied by the same number.

b. Reducing a fraction: The same number must be removed from the numerator and denominator.

REVIEW

117. $\$3{,}575 - \$235 - \$782$
$-\$148 - \$103 = \$2{,}307$

Section 3.2: Multiplying Fractions

VOCABULARY

1. When a fraction is followed by the word *of*, such as $\dfrac{1}{3}$ *of*, it indicates that we are to find a part of some quantity using <u>multiplication</u>.

3. To <u>simplify</u> a fraction we remove the common factors of the numerator and denominator.

5. The <u>area</u> of a triangle is the amount of surface that it encloses.

CONCEPTS

7. To multiply two fractions, multiply the <u>numerators</u> and multiply the <u>denominators</u>. Then <u>simplify</u> if possible.

9. a. negative

b. positive

c. positive

d. negative

11. Area of a triangle = $\dfrac{1}{2}$ (base $\times$ height) or

$A = \dfrac{1}{2}bh$

NOTATION

13. a. $4 = \dfrac{4}{1}$ b. $-3 = -\dfrac{3}{1}$

15. $\dfrac{5}{8} \cdot \dfrac{7}{15} = \dfrac{5 \cdot 7}{8 \cdot 15}$

$= \dfrac{5 \cdot 7}{2 \cdot 2 \cdot 2 \cdot 3 \cdot 5}$

$= \dfrac{\cancel{5} \cdot 7}{2 \cdot 2 \cdot 2 \cdot 3 \cdot \cancel{5}}$

$= \dfrac{7}{24}$

GUIDED PRACTICE

17. $\dfrac{1}{4} \cdot \dfrac{1}{2} = \dfrac{1}{8}$

19. $\dfrac{1}{9} \cdot \dfrac{1}{5} = \dfrac{1}{45}$

21. $\dfrac{2}{3} \cdot \dfrac{7}{9} = \dfrac{14}{27}$

23. $\dfrac{8}{11} \cdot \dfrac{3}{7} = \dfrac{24}{77}$

25. $-\dfrac{4}{5} \cdot \dfrac{1}{3} = -\dfrac{4}{15}$

27. $\dfrac{5}{6}\left(-\dfrac{7}{12}\right) = -\dfrac{35}{72}$

29. $\dfrac{1}{8} \cdot 9 = \dfrac{1}{8} \cdot \dfrac{9}{1} = \dfrac{9}{8}$

31. $\dfrac{1}{2} \cdot 5 = \dfrac{1}{2} \cdot \dfrac{5}{1} = \dfrac{5}{2}$

33. $\dfrac{11}{10} \cdot \dfrac{5}{11} = \dfrac{11}{2 \cdot 5} \cdot \dfrac{5}{11} = \dfrac{\cancel{11}}{2 \cdot \cancel{5}} \cdot \dfrac{\cancel{5}}{\cancel{11}} = \dfrac{1}{2}$

35. $\dfrac{6}{49} \cdot \dfrac{7}{6} = \dfrac{6}{7 \cdot 7} \cdot \dfrac{7}{6} = \dfrac{\cancel{6}}{7 \cdot \cancel{7}} \cdot \dfrac{\cancel{7}}{\cancel{6}} = \dfrac{1}{7}$

37. $\dfrac{3}{4}\left(-\dfrac{8}{35}\right)\left(-\dfrac{7}{12}\right) = \dfrac{3}{4}\left(-\dfrac{2 \cdot 4}{5 \cdot 7}\right)\left(-\dfrac{7}{2 \cdot 2 \cdot 3}\right)$
$= \dfrac{\cancel{3}}{\cancel{4}}\left(-\dfrac{\cancel{2} \cdot \cancel{4}}{5 \cdot \cancel{7}}\right)\left(-\dfrac{\cancel{7}}{\cancel{2} \cdot 2 \cdot \cancel{3}}\right) = \dfrac{1}{10}$

39. $-\dfrac{5}{8}\left(\dfrac{16}{27}\right)\left(-\dfrac{9}{25}\right) = -\dfrac{5}{8}\left(\dfrac{2 \cdot 8}{3 \cdot 9}\right)\left(-\dfrac{9}{5 \cdot 5}\right)$
$= -\dfrac{\cancel{5}}{\cancel{8}}\left(\dfrac{2 \cdot \cancel{8}}{3 \cdot \cancel{9}}\right)\left(-\dfrac{\cancel{9}}{5 \cdot \cancel{5}}\right) = \dfrac{2}{15}$

41. a. $\left(\dfrac{3}{5}\right)^2 = \dfrac{3}{5} \cdot \dfrac{3}{5} = \dfrac{9}{25}$

b. $\left(-\dfrac{3}{5}\right)^2 = \left(-\dfrac{3}{5}\right)\left(-\dfrac{3}{5}\right) = \dfrac{9}{25}$

43. a. $-\left(-\dfrac{1}{6}\right)^2 = -\left(-\dfrac{1}{6}\right)\left(-\dfrac{1}{6}\right) = -\dfrac{1}{36}$

b. $\left(-\dfrac{1}{6}\right)^3 = \left(-\dfrac{1}{6}\right)\left(-\dfrac{1}{6}\right)\left(-\dfrac{1}{6}\right) = -\dfrac{1}{216}$

45. $\dfrac{3}{4}$ of $\dfrac{5}{8} = \dfrac{3}{4} \cdot \dfrac{5}{8} = \dfrac{15}{32}$

47. $\dfrac{1}{6}$ of $54 = \dfrac{1}{6} \cdot \dfrac{54}{1} = \dfrac{54}{6} = 9$

49. $\dfrac{1}{2} \cdot 10 \cdot 3 = 5 \cdot 3 = 15 \, ft^2$

51. $\dfrac{1}{2} \cdot 18 \cdot 7 = 9 \cdot 7 = 63 \, in^2$

53. $\dfrac{1}{2} \cdot 4 \cdot 3 = 2 \cdot 3 = 6 \, m^2$

55. $\dfrac{1}{2} \cdot 24 \cdot 5 = 12 \cdot 5 = 60 \, ft^2$

TRY IT YOURSELF

57.

·	$\dfrac{1}{2}$	$\dfrac{1}{3}$	$\dfrac{1}{4}$	$\dfrac{1}{5}$	$\dfrac{1}{6}$
$\dfrac{1}{2}$	$\dfrac{1}{4}$	$\dfrac{1}{6}$	$\dfrac{1}{8}$	$\dfrac{1}{10}$	$\dfrac{1}{12}$
$\dfrac{1}{3}$	$\dfrac{1}{6}$	$\dfrac{1}{9}$	$\dfrac{1}{12}$	$\dfrac{1}{15}$	$\dfrac{1}{18}$
$\dfrac{1}{4}$	$\dfrac{1}{8}$	$\dfrac{1}{12}$	$\dfrac{1}{16}$	$\dfrac{1}{20}$	$\dfrac{1}{24}$
$\dfrac{1}{5}$	$\dfrac{1}{10}$	$\dfrac{1}{15}$	$\dfrac{1}{20}$	$\dfrac{1}{25}$	$\dfrac{1}{30}$
$\dfrac{1}{6}$	$\dfrac{1}{12}$	$\dfrac{1}{18}$	$\dfrac{1}{24}$	$\dfrac{1}{30}$	$\dfrac{1}{36}$

59. $-\dfrac{15}{24} \cdot \dfrac{8}{25} = -\dfrac{3 \cdot 5}{3 \cdot 8} \cdot \dfrac{8}{5 \cdot 5} = -\dfrac{\cancel{3} \cdot \cancel{5}}{\cancel{3} \cdot \cancel{8}} \cdot \dfrac{\cancel{8}}{\cancel{5} \cdot 5} = -\dfrac{1}{5}$

61. $\dfrac{3}{8} \cdot \dfrac{7}{16} = \dfrac{21}{128}$

63. $\left(\dfrac{2}{3}\right)\left(-\dfrac{1}{16}\right)\left(-\dfrac{4}{5}\right) = \left(\dfrac{2}{3}\right)\left(-\dfrac{1}{2\cdot 2\cdot 4}\right)\left(-\dfrac{4}{5}\right)$

$= \left(\dfrac{\cancel{2}}{3}\right)\left(-\dfrac{1}{\cancel{2}\cdot 2\cdot \cancel{4}}\right)\left(-\dfrac{\cancel{4}}{5}\right)$

$= \dfrac{1}{30}$

65. $-\dfrac{5}{6}\cdot 18 = -\dfrac{5}{6}\cdot\dfrac{3\cdot 6}{1} = -\dfrac{5}{\cancel{6}}\cdot\dfrac{3\cdot\cancel{6}}{1} = -15$

67. $\left(-\dfrac{3}{4}\right)^3 = \left(-\dfrac{3}{4}\right)\left(-\dfrac{3}{4}\right)\left(-\dfrac{3}{4}\right) = -\dfrac{27}{64}$

69. $\dfrac{3}{4}\cdot\dfrac{4}{3} = \dfrac{12}{12} = 1$

71. $\dfrac{5}{3}\left(-\dfrac{6}{15}\right)(-4) = \dfrac{5}{3}\left(-\dfrac{2\cdot 3}{3\cdot 5}\right)(-4)$

$= \dfrac{\cancel{5}}{\cancel{3}}\left(-\dfrac{2\cdot\cancel{3}}{3\cdot\cancel{5}}\right)(-4) = \dfrac{8}{3}$

73. $-\dfrac{11}{12}\cdot\dfrac{18}{55}\cdot 5 = -\dfrac{11}{2\cdot 6}\cdot\dfrac{3\cdot 6}{5\cdot 11}\cdot\dfrac{5}{1}$

$= -\dfrac{\cancel{11}}{2\cdot\cancel{6}}\cdot\dfrac{3\cdot\cancel{6}}{\cancel{5}\cdot\cancel{11}}\cdot\dfrac{\cancel{5}}{1} = -\dfrac{3}{2}$

75. $\left(-\dfrac{11}{21}\right)\left(-\dfrac{14}{33}\right) = \left(-\dfrac{11}{3\cdot 7}\right)\left(-\dfrac{2\cdot 7}{3\cdot 11}\right)$

$= \left(-\dfrac{\cancel{11}}{3\cdot\cancel{7}}\right)\left(-\dfrac{2\cdot\cancel{7}}{3\cdot\cancel{11}}\right) = \dfrac{2}{9}$

77. $-\left(-\dfrac{5}{9}\right)^2 = -\left(-\dfrac{5}{9}\right)\left(-\dfrac{5}{9}\right) = -\dfrac{25}{81}$

79. $\dfrac{7}{10}\left(\dfrac{20}{21}\right) = \dfrac{7}{10}\left(\dfrac{2\cdot 10}{3\cdot 7}\right)$

$= \dfrac{\cancel{7}}{\cancel{10}}\left(\dfrac{2\cdot\cancel{10}}{3\cdot\cancel{7}}\right) = \dfrac{2}{3}$

81. $\dfrac{3}{4}\left(\dfrac{5}{7}\right)\left(\dfrac{2}{3}\right)\left(\dfrac{7}{3}\right) = \dfrac{3}{2\cdot 2}\left(\dfrac{5}{7}\right)\left(\dfrac{2}{3}\right)\left(\dfrac{7}{3}\right)$

$= \dfrac{\cancel{3}}{\cancel{2}\cdot 2}\left(\dfrac{5}{\cancel{7}}\right)\left(\dfrac{\cancel{2}}{\cancel{3}}\right)\left(\dfrac{\cancel{7}}{3}\right) = \dfrac{5}{6}$

83. $-\dfrac{14}{15}\left(-\dfrac{11}{8}\right) = \left(-\dfrac{2\cdot 7}{15}\right)\left(-\dfrac{11}{2\cdot 4}\right)$

$= \left(-\dfrac{\cancel{2}\cdot 7}{15}\right)\left(-\dfrac{11}{\cancel{2}\cdot 4}\right) = \dfrac{77}{60}$

85. $\dfrac{3}{16}\cdot 4\cdot\dfrac{2}{3} = \dfrac{3\cdot 4\cdot 2}{3\cdot 16} = \dfrac{24}{48} = \dfrac{\cancel{24}}{2\cdot\cancel{24}} = \dfrac{1}{2}$

APPLICATIONS

87. $\dfrac{3}{5}$ of $100 = \dfrac{3}{5}\cdot 100 = \dfrac{3}{\cancel{5}}\cdot\dfrac{\cancel{5}\cdot 20}{1} = 60$ votes

89. TENNIS BALLS

h_1 = height of first bounce

h_2 = height of second bounce

h_3 = height of third bounce

$h_1 = \dfrac{1}{3}(54)$

$h_1 = \dfrac{1}{3}\left(\dfrac{54}{1}\right)$

$h_1 = \dfrac{1}{\cancel{3}}\left(\dfrac{18\cdot\cancel{3}}{1}\right) = \dfrac{18}{1} = 18$

$h_2 = \dfrac{1}{3}(18)$

$h_2 = \dfrac{1}{3}\left(\dfrac{18}{1}\right)$

$h_2 = \dfrac{1}{\cancel{3}}\left(\dfrac{6\cdot\cancel{3}}{1}\right) = \dfrac{6}{1} = 6$

$$h_3 = \frac{1}{3}(6)$$

$$h_3 = \frac{1}{3}\left(\frac{6}{1}\right)$$

$$h_3 = \frac{1}{\cancel{3}}\left(\frac{2\cdot\cancel{3}}{1}\right) = \frac{2}{1} = 2$$

18 inches on first bounce, 6 inches on second bounce, and 2 inches on third bounce.

91. $\frac{1}{2}\cdot\frac{3}{4} = \frac{3}{8}$ cups sugar

$\frac{1}{2}\cdot\frac{1}{3} = \frac{1}{6}$ cups molasses

93.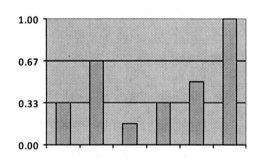

95. $\frac{1}{2}\cdot 6\cdot 9 = 3\cdot 9 = 27\,ft^2$

97. $\frac{1}{2}\cdot 12\cdot 7 = 6\cdot 7 = 42\,ft^2$

99. $\frac{1}{2}\cdot 106\cdot 182 = 53\cdot 182 = 9{,}646\,mi^2$

101. $12\cdot\frac{1}{16} = \frac{3\cdot\cancel{4}}{1}\cdot\frac{1}{4\cdot\cancel{4}} = \frac{3}{4}$ inches

WRITING

103. $\frac{1}{2}$ of a pizza, $\frac{2}{3}$ of an inning, $\frac{1}{4}$ of an hour

105. It means more than half the class voted to postpone – for example $\frac{13}{22}$ of the class.

107. The student neglected to factor and cancel before multiplying.

REVIEW

109. $\frac{-8}{4} = \frac{-2\cdot\cancel{4}}{\cancel{4}} = -2$

111. $-736\div(-32) = \frac{-736}{-32} = \frac{-23\cdot 32}{-32} = 23$

Section 3.3: Dividing Fractions

VOCABULARY

1. The reciprocal of $\frac{5}{12}$ is $\frac{12}{5}$.

3. The answer to a division is called the quotient.

CONCEPTS

5. a. To divide two fractions, multiply the first fraction by the reciprocal of the second fraction.

b. $\frac{1}{2}\div\frac{2}{3} = \frac{1}{2}\cdot\frac{3}{2}$

7. a. Negative

b. Positive

9. a. $\frac{4}{5}\cdot\frac{5}{4} = \frac{20}{20} = 1$

b. $\left(-\frac{3}{5}\right)\left(-\frac{5}{3}\right) = \frac{15}{15} = 1$

NOTATION

11. $\frac{4}{9}\div\frac{8}{27} = \frac{4}{9}\cdot\frac{27}{8}$

$= \frac{4\cdot 27}{9\cdot 8}$

$= \frac{\cancel{4}\cdot 3\cdot\cancel{9}}{\cancel{9}\cdot 2\cdot\cancel{4}}$

$= \frac{3}{2}$

GUIDED PRACTICE

13. a. $\dfrac{7}{6}$ b. $-\dfrac{8}{15}$ c. $\dfrac{1}{10}$

15. a. $\dfrac{8}{11}$ b. -14 c. $-\dfrac{1}{63}$

17. $\dfrac{1}{8} \div \dfrac{2}{3} = \dfrac{1}{8} \cdot \dfrac{3}{2} = \dfrac{3}{16}$

19. $\dfrac{2}{23} \div \dfrac{1}{7} = \dfrac{2}{23} \cdot \dfrac{7}{1} = \dfrac{14}{23}$

21. $\dfrac{25}{32} \div \dfrac{5}{28} = \dfrac{25}{32} \cdot \dfrac{28}{5}$

    $= \dfrac{\cancel{5} \cdot 5}{\cancel{4} \cdot 8} \cdot \dfrac{\cancel{4} \cdot 7}{\cancel{5}} = \dfrac{35}{8}$

23. $\dfrac{27}{32} \div \dfrac{9}{8} = \dfrac{27}{32} \cdot \dfrac{8}{9}$

    $= \dfrac{3 \cdot \cancel{9}}{4 \cdot \cancel{8}} \cdot \dfrac{\cancel{8}}{\cancel{9}} = \dfrac{3}{4}$

25. $50 \div \dfrac{10}{9} = \dfrac{50}{1} \cdot \dfrac{9}{10}$

    $= \dfrac{5 \cdot \cancel{10}}{1} \cdot \dfrac{9}{\cancel{10}} = 45$

27. $150 \div \dfrac{15}{32} = \dfrac{150}{1} \cdot \dfrac{32}{15}$

    $= \dfrac{\cancel{15} \cdot 10}{1} \cdot \dfrac{32}{\cancel{15}} = 320$

29. $\dfrac{1}{8} \div \left(-\dfrac{1}{32}\right) = \dfrac{1}{8} \cdot \left(-\dfrac{32}{1}\right)$

    $= \dfrac{-32}{8} = -4$

31. $\dfrac{2}{5} \div \left(-\dfrac{4}{35}\right) = \dfrac{2}{5} \cdot \left(-\dfrac{35}{4}\right)$

    $= \dfrac{\cancel{2}}{\cancel{5}} \left(-\dfrac{\cancel{5} \cdot 7}{\cancel{2} \cdot 2}\right) = -\dfrac{7}{2}$

33. $-\dfrac{28}{55} \div (-7) = -\dfrac{28}{55} \cdot \left(-\dfrac{1}{7}\right)$

    $= -\dfrac{4 \cdot \cancel{7}}{55} \cdot \left(-\dfrac{1}{\cancel{7}}\right) = \dfrac{4}{55}$

35. $-\dfrac{33}{23} \div (-11) = -\dfrac{33}{23} \cdot \left(-\dfrac{1}{11}\right)$

    $= -\dfrac{3 \cdot \cancel{11}}{23} \cdot \left(-\dfrac{1}{\cancel{11}}\right) = \dfrac{3}{23}$

TRY IT YOURSELF

37. $120 \div \dfrac{12}{5} = \dfrac{120}{1} \cdot \dfrac{5}{12}$

    $= \dfrac{10 \cdot \cancel{12}}{1} \cdot \dfrac{5}{\cancel{12}} = 50$

39. $\dfrac{1}{2} \div \dfrac{3}{5} = \dfrac{1}{2} \cdot \dfrac{5}{3} = \dfrac{5}{6}$

41. $\left(-\dfrac{7}{4}\right) \div \left(-\dfrac{21}{8}\right) = \left(-\dfrac{7}{4}\right)\left(-\dfrac{8}{21}\right)$

    $= \left(-\dfrac{\cancel{7}}{\cancel{4}}\right)\left(-\dfrac{2 \cdot \cancel{4}}{3 \cdot \cancel{7}}\right) = \dfrac{2}{3}$

43. $\dfrac{4}{5} \div \dfrac{4}{5} = \dfrac{4}{5} \cdot \dfrac{5}{4} = \dfrac{20}{20} = 1$

45. $-\dfrac{15}{32} \div \dfrac{3}{4} = -\dfrac{15}{32} \cdot \dfrac{4}{3}$

    $= -\dfrac{\cancel{3} \cdot 5}{\cancel{4} \cdot 8} \cdot \dfrac{\cancel{4}}{\cancel{3}} = -\dfrac{5}{8}$

47. $3 \div \dfrac{1}{12} = \dfrac{3}{1} \cdot \dfrac{12}{1} = 36$

49. $-\dfrac{4}{5} \div (-6) = -\dfrac{4}{5} \left(-\dfrac{1}{6}\right)$

    $= -\dfrac{2 \cdot \cancel{2}}{5} \cdot \left(-\dfrac{1}{\cancel{2} \cdot 3}\right) = \dfrac{2}{15}$

51. $\dfrac{15}{16} \div 180 = \dfrac{15}{16} \cdot \dfrac{1}{180}$

$= \dfrac{\cancel{15}}{16} \cdot \dfrac{1}{12 \cdot \cancel{15}} = \dfrac{1}{192}$

53. $-\dfrac{9}{10} \div \dfrac{4}{15} = -\dfrac{9}{10} \cdot \dfrac{15}{4}$

$= -\dfrac{9}{2 \cdot \cancel{5}} \cdot \dfrac{3 \cdot \cancel{5}}{4} = -\dfrac{27}{8}$

55. $\dfrac{9}{10} \div \left(-\dfrac{3}{25}\right) = \dfrac{9}{10} \cdot \left(-\dfrac{25}{3}\right)$

$= \dfrac{3 \cdot \cancel{3}}{2 \cdot \cancel{5}} \cdot \left(-\dfrac{\cancel{5} \cdot 5}{\cancel{3}}\right) = -\dfrac{15}{2}$

57. $\dfrac{3}{16} \div \dfrac{1}{9} = \dfrac{3}{16} \cdot \dfrac{9}{1} = \dfrac{27}{16}$

59. $-\dfrac{1}{8} \div 8 = -\dfrac{1}{8} \cdot \dfrac{1}{8} = -\dfrac{1}{64}$

61. $\dfrac{7}{6} \cdot \dfrac{9}{49} = \dfrac{\cancel{7}}{2 \cdot \cancel{3}} \cdot \dfrac{\cancel{3} \cdot 3}{\cancel{7} \cdot 7} = \dfrac{3}{14}$

63. $-\dfrac{4}{5} \div \left(-\dfrac{3}{2}\right) = -\dfrac{4}{5} \cdot \left(-\dfrac{2}{3}\right) = \dfrac{8}{15}$

65. $\dfrac{13}{16} \div 2 = \dfrac{13}{16} \cdot \dfrac{1}{2} = \dfrac{13}{32}$

67. $\left(-\dfrac{11}{21}\right)\left(-\dfrac{14}{33}\right)$

$= \left(-\dfrac{\cancel{11}}{3 \cdot \cancel{7}}\right)\left(-\dfrac{2 \cdot \cancel{7}}{3 \cdot \cancel{11}}\right) = \dfrac{2}{9}$

69. $-\dfrac{15}{32} \div \dfrac{5}{64} = -\dfrac{15}{32} \cdot \dfrac{64}{5}$

$= -\dfrac{3 \cdot \cancel{5}}{\cancel{32}} \cdot \dfrac{2 \cdot \cancel{32}}{\cancel{5}} = -6$

71. $11 \cdot \dfrac{1}{6} = \dfrac{11}{1} \cdot \dfrac{1}{6} = \dfrac{11}{6}$

73. $\dfrac{3}{4} \cdot \dfrac{5}{7} = \dfrac{15}{28}$

75. $\dfrac{25}{7} \div \left(-\dfrac{30}{21}\right) = \dfrac{25}{7} \cdot \left(-\dfrac{21}{30}\right)$

$= \dfrac{\cancel{5} \cdot 5}{\cancel{7}} \cdot \left(-\dfrac{\cancel{3} \cdot \cancel{7}}{2 \cdot \cancel{3} \cdot \cancel{5}}\right) = -\dfrac{5}{2}$

APPLICATIONS

77. $\dfrac{3}{8} \div \dfrac{3}{32} = \dfrac{3}{8} \cdot \dfrac{32}{3} = \dfrac{\cancel{3}}{\cancel{8}} \cdot \dfrac{4 \cdot \cancel{8}}{\cancel{3}} = 4$ coats.

79. $\dfrac{3}{4} \div \dfrac{1}{8} = \dfrac{3}{4} \cdot \dfrac{8}{1} = \dfrac{24}{4} = 6$ cups.

81. a. $12 \div \dfrac{2}{5} = \dfrac{12}{1} \cdot \dfrac{5}{2} = \dfrac{6}{1} \cdot \dfrac{5}{1} = 30$ days

   b. $7 + 8 = 15$ miles

   c. $15 \div \dfrac{3}{5} = \dfrac{15}{1} \cdot \dfrac{5}{3} = \dfrac{5}{1} \cdot \dfrac{5}{1} = 25$ days

   d. Route 2 will take less time.

83. a. 16

   b. $\dfrac{3}{4}$ in

   c. $\dfrac{3}{4} \div 90 = \dfrac{3}{4} \cdot \dfrac{1}{90} = \dfrac{1}{120}$ in

85. $6,284 \div \dfrac{4}{5} = \dfrac{6,284}{1} \cdot \dfrac{5}{4} = \dfrac{\cancel{4} \cdot 1571}{1} \cdot \dfrac{5}{\cancel{4}} = 7,855$ sections.

WRITING

87. To divide two fractions, multiply the first fraction by the reciprocal of the second fraction.

89. 0 does not have a reciprocal because $\dfrac{1}{0}$ is undefined.

91. Bill has 10 pizzas. If he evenly slices them so that each guest receives $\frac{1}{5}$ of a pizza, how many guests does Bill have?

REVIEW

93. The symbol < means is less than.

95. Zero is neither positive nor negative.

97.

Section 3.4: Adding and Subtracting Fractions

VOCABULARY

1. Because the denominators of $\frac{3}{8}$ and $\frac{7}{8}$ are the same number, we say that they have a common denominator.

3. To build an equivalent fraction with a denominator of 18 we multiply $\frac{4}{9}$ by a 1 in the form of $\frac{2}{2}$.

CONCEPTS

5. To add (or subtract) fractions that have the same denominator, add (or subtract) their numerators and write the sum (or difference) over the common denominator. Simplify the result, if possible.

7. When adding (or subtracting) two fractions with different denominators, if the smaller denominator is a factor of the larger denominator, the larger denominator is the LCD.

9. Since $4 \cdot 9 = 36$, the form of 1 should be $\frac{9}{9}$.

11. a. a 5 appears once

    b. a 3 appears twice

    c. a 2 appears three times

13. $\left. \begin{array}{l} 20 = 2 \cdot 2 \cdot 5 \\ 30 = 2 \cdot 3 \cdot 5 \\ 90 = 2 \cdot 3 \cdot 3 \cdot 5 \end{array} \right\} LCD = 2 \cdot 2 \cdot 3 \cdot 3 \cdot 5 = 180$

NOTATION

15. $\frac{2}{5} + \frac{1}{7} = \frac{2}{5} \cdot \frac{7}{7} + \frac{1}{7} \cdot \frac{5}{5}$

    $= \frac{14}{35} + \frac{5}{35}$

    $= \frac{14+5}{35}$

    $= \frac{19}{35}$

GUIDED PRACTICE

17. $\frac{4}{9} + \frac{1}{9} = \frac{5}{9}$

19. $\frac{3}{8} + \frac{1}{8} = \frac{4}{8} = \frac{\cancel{4}}{2 \cdot \cancel{4}} = \frac{1}{2}$

21. $\frac{11}{15} - \frac{7}{15} = \frac{4}{15}$

23. $\frac{11}{20} - \frac{3}{20} = \frac{8}{20} = \frac{2 \cdot \cancel{4}}{\cancel{4} \cdot 5} = \frac{2}{5}$

25. $-\frac{11}{5} - \left(-\frac{8}{5}\right) = \frac{-11}{5} + \frac{8}{5}$

    $= \frac{-3}{5} = -\frac{3}{5}$

27. $-\frac{7}{21} - \left(-\frac{2}{21}\right) = \frac{-7}{21} + \frac{2}{21}$

    $= \frac{-5}{21} = -\frac{5}{21}$

29. $\frac{19}{40} - \frac{3}{40} - \frac{1}{40} = \frac{15}{40}$

    $= \frac{3 \cdot \cancel{5}}{\cancel{5} \cdot 8} = \frac{3}{8}$

31. $\dfrac{13}{33} + \dfrac{1}{33} + \dfrac{7}{33} = \dfrac{21}{33}$

$= \dfrac{\cancel{3} \cdot 7}{\cancel{3} \cdot 11} = \dfrac{7}{11}$

33. $\dfrac{1}{3} + \dfrac{1}{7} = \dfrac{1}{3} \cdot \dfrac{7}{7} + \dfrac{1}{7} \cdot \dfrac{3}{3}$

$= \dfrac{7}{21} + \dfrac{3}{21} = \dfrac{10}{21}$

35. $\dfrac{2}{5} + \dfrac{1}{2} = \dfrac{2}{5} \cdot \dfrac{2}{2} + \dfrac{1}{2} \cdot \dfrac{5}{5}$

$= \dfrac{4}{10} + \dfrac{5}{10} = \dfrac{9}{10}$

37. $\dfrac{4}{5} - \dfrac{3}{4} = \dfrac{4}{5} \cdot \dfrac{4}{4} - \dfrac{3}{4} \cdot \dfrac{5}{5}$

$= \dfrac{16}{20} - \dfrac{15}{20} = \dfrac{1}{20}$

39. $\dfrac{3}{4} - \dfrac{2}{7} = \dfrac{3}{4} \cdot \dfrac{7}{7} - \dfrac{2}{7} \cdot \dfrac{4}{4}$

$= \dfrac{21}{28} - \dfrac{8}{28} = \dfrac{13}{28}$

41. $\dfrac{11}{12} - \dfrac{2}{3} = \dfrac{11}{12} - \dfrac{2}{3} \cdot \dfrac{4}{4}$

$= \dfrac{11}{12} - \dfrac{8}{12} = \dfrac{3}{12} = \dfrac{3}{4 \cdot 3} = \dfrac{1}{4}$

43. $\dfrac{9}{14} - \dfrac{1}{7} = \dfrac{9}{14} - \dfrac{1}{7} \cdot \dfrac{2}{2}$

$= \dfrac{9}{14} - \dfrac{2}{14} = \dfrac{7}{14} = \dfrac{\cancel{7}}{2 \cdot \cancel{7}}$

$= \dfrac{1}{2}$

45. $-2 + \dfrac{5}{9} = \dfrac{-2}{1} + \dfrac{5}{9}$

$= \dfrac{-2}{1} \cdot \dfrac{9}{9} + \dfrac{5}{9} = \dfrac{-18}{9} + \dfrac{5}{9}$

$= \dfrac{-13}{9} = -\dfrac{13}{9}$

47. $-3 + \dfrac{9}{4} = \dfrac{-3}{1} + \dfrac{9}{4}$

$= \dfrac{-3}{1} \cdot \dfrac{4}{4} + \dfrac{9}{4} = \dfrac{-12}{4} + \dfrac{9}{4}$

$= \dfrac{-3}{4} = -\dfrac{3}{4}$

49. $\dfrac{1}{6} + \dfrac{5}{8} = \dfrac{1}{6} \cdot \dfrac{4}{4} + \dfrac{5}{8} \cdot \dfrac{3}{3}$

$= \dfrac{4}{24} + \dfrac{15}{24} = \dfrac{19}{24}$

51. $\dfrac{4}{9} + \dfrac{5}{12} = \dfrac{4}{9} \cdot \dfrac{4}{4} + \dfrac{5}{12} \cdot \dfrac{3}{3}$

$= \dfrac{16}{36} + \dfrac{15}{36} = \dfrac{31}{36}$

53. $\dfrac{9}{10} - \dfrac{3}{14} = \dfrac{9}{10} \cdot \dfrac{7}{7} - \dfrac{3}{14} \cdot \dfrac{5}{5}$

$= \dfrac{63}{70} - \dfrac{15}{70} = \dfrac{48}{70} =$

$\dfrac{\cancel{2} \cdot 24}{\cancel{2} \cdot 35} = \dfrac{24}{35}$

55. $\dfrac{11}{12} - \dfrac{7}{15} = \dfrac{11}{12} \cdot \dfrac{5}{5} - \dfrac{7}{15} \cdot \dfrac{4}{4}$

$= \dfrac{55}{60} - \dfrac{28}{60} = \dfrac{27}{60}$

$= \dfrac{\cancel{3} \cdot 9}{\cancel{3} \cdot 20} = \dfrac{9}{20}$

57. $\dfrac{3}{8} \;\square\; \dfrac{5}{16}$

$\dfrac{3}{8} \cdot \dfrac{2}{2} \;\square\; \dfrac{5}{16}$

$\dfrac{6}{16} \;\boxed{>}\; \dfrac{5}{16}$

$\dfrac{3}{8}$

59. $\dfrac{4}{5} \square \dfrac{2}{3}$

$\dfrac{4}{5} \cdot \dfrac{3}{3} \square \dfrac{2}{3} \cdot \dfrac{5}{5}$

$\dfrac{12}{15} \geq \dfrac{10}{15}$

$\dfrac{4}{5}$

61. $\dfrac{7}{9} \square \dfrac{11}{12}$

$\dfrac{7}{9} \cdot \dfrac{4}{4} \square \dfrac{11}{12} \cdot \dfrac{3}{3}$

$\dfrac{28}{36} \leq \dfrac{33}{36}$

$\dfrac{11}{12}$

63. $\dfrac{23}{20} \square \dfrac{7}{6}$

$\dfrac{23}{20} \cdot \dfrac{3}{3} \square \dfrac{7}{6} \cdot \dfrac{10}{10}$

$\dfrac{69}{60} \leq \dfrac{70}{60}$

$\dfrac{7}{6}$

65. $\dfrac{1}{6} + \dfrac{5}{18} + \dfrac{2}{9} = \dfrac{1}{6} \cdot \dfrac{3}{3} + \dfrac{5}{18} + \dfrac{2}{9} \cdot \dfrac{2}{2}$

$= \dfrac{3}{18} + \dfrac{5}{18} + \dfrac{4}{18} = \dfrac{12}{18} = \dfrac{2 \cdot \cancel{6}}{3 \cdot \cancel{6}}$

$= \dfrac{2}{3}$

67. $\dfrac{4}{15} + \dfrac{2}{3} + \dfrac{1}{6} = \dfrac{4}{15} \cdot \dfrac{2}{2} + \dfrac{2}{3} \cdot \dfrac{10}{10} + \dfrac{1}{6} \cdot \dfrac{5}{5}$

$= \dfrac{8}{30} + \dfrac{20}{30} + \dfrac{5}{30} = \dfrac{33}{30} = \dfrac{\cancel{3} \cdot 11}{\cancel{3} \cdot 10} = \dfrac{11}{10}$

TRY IT YOURSELF

69. $-\dfrac{1}{12} - \left(-\dfrac{5}{12}\right) = \dfrac{-1}{12} + \dfrac{5}{12}$

$= \dfrac{4}{12} = \dfrac{\cancel{4}}{3 \cdot \cancel{4}} = \dfrac{1}{3}$

71. $\dfrac{4}{5} + \dfrac{2}{3} = \dfrac{4}{5} \cdot \dfrac{3}{3} + \dfrac{2}{3} \cdot \dfrac{5}{5}$

$= \dfrac{12}{15} + \dfrac{10}{15} = \dfrac{22}{15}$

73. $\dfrac{12}{25} - \dfrac{1}{25} - \dfrac{1}{25} = \dfrac{10}{25}$

$= \dfrac{2 \cdot \cancel{5}}{5 \cdot \cancel{5}} = \dfrac{2}{5}$

75. $-\dfrac{7}{20} - \dfrac{1}{5} = -\dfrac{7}{20} - \dfrac{1}{5} \cdot \dfrac{4}{4}$

$= \dfrac{-7}{20} - \dfrac{4}{20} = \dfrac{-11}{20} = -\dfrac{11}{20}$

77. $3\dfrac{1}{16} \cdot 4\dfrac{4}{7} = \dfrac{49}{16} \cdot \dfrac{32}{7} = \dfrac{\cancel{7} \cdot 7}{\cancel{16}} \cdot \dfrac{2 \cdot \cancel{16}}{\cancel{7}} = 14$

79. $\dfrac{11}{12} - \dfrac{2}{3} = \dfrac{11}{12} - \dfrac{2}{3} \cdot \dfrac{4}{4}$

$= \dfrac{11}{12} - \dfrac{8}{12} = \dfrac{3}{12}$

$= \dfrac{\cancel{3}}{\cancel{3} \cdot 4} = \dfrac{1}{4}$

81. $\dfrac{2}{3} + \dfrac{4}{5} + \dfrac{5}{6} = \dfrac{2}{3} \cdot \dfrac{10}{10} + \dfrac{4}{5} \cdot \dfrac{6}{6} + \dfrac{5}{6} \cdot \dfrac{5}{5}$

$= \dfrac{20}{30} + \dfrac{24}{30} + \dfrac{25}{30} = \dfrac{69}{30} = \dfrac{\cancel{3} \cdot 23}{\cancel{3} \cdot 10}$

$= \dfrac{23}{10}$

83. $\dfrac{9}{20} - \dfrac{1}{30} = \dfrac{9}{20} \cdot \dfrac{3}{3} - \dfrac{1}{30} \cdot \dfrac{2}{2}$

$= \dfrac{27}{60} - \dfrac{2}{60} = \dfrac{25}{60} = \dfrac{\cancel{5} \cdot 5}{\cancel{5} \cdot 12}$

$= \dfrac{5}{12}$

85. $\dfrac{27}{50} + \dfrac{5}{16} = \dfrac{27}{50} \cdot \dfrac{8}{8} + \dfrac{5}{16} \cdot \dfrac{25}{25}$

$= \dfrac{216}{400} + \dfrac{125}{400} = \dfrac{341}{400}$

87. $\dfrac{13}{20} - \dfrac{1}{5} = \dfrac{13}{20} - \dfrac{1}{5} \cdot \dfrac{4}{4}$

$= \dfrac{13}{20} - \dfrac{4}{20} = \dfrac{9}{20}$

89. $\dfrac{37}{103} - \dfrac{17}{103} = \dfrac{20}{103}$

91. $-\dfrac{3}{4} - 5 = \dfrac{-3}{4} - \dfrac{5}{1} \cdot \dfrac{4}{4}$

$= \dfrac{-3}{4} - \dfrac{20}{4} = \dfrac{-23}{4} = -\dfrac{23}{4}$

93. $\dfrac{4}{27} + \dfrac{1}{6} = \dfrac{4}{27} \cdot \dfrac{2}{2} + \dfrac{1}{6} \cdot \dfrac{9}{9}$

$= \dfrac{8}{54} + \dfrac{9}{54} = \dfrac{17}{54}$

95. $\dfrac{7}{30} - \dfrac{19}{75} = \dfrac{7}{30} \cdot \dfrac{5}{5} - \dfrac{19}{75} \cdot \dfrac{2}{2}$

$= \dfrac{35}{150} - \dfrac{38}{150} = \dfrac{-3}{150}$

$= \dfrac{-\cancel{3}}{\cancel{3} \cdot 50} = \dfrac{-1}{50} = -\dfrac{1}{50}$

97. $\dfrac{11}{60} - \dfrac{2}{45} = \dfrac{11}{60} \cdot \dfrac{3}{3} - \dfrac{2}{45} \cdot \dfrac{4}{4}$

$= \dfrac{33}{180} - \dfrac{8}{180} = \dfrac{25}{180}$

$= \dfrac{\cancel{5} \cdot 5}{\cancel{5} \cdot 36} = \dfrac{5}{36}$

99. $\dfrac{2}{15} - \dfrac{5}{12} = \dfrac{2}{15} \cdot \dfrac{4}{4} - \dfrac{5}{12} \cdot \dfrac{5}{5}$

$= \dfrac{8}{60} - \dfrac{25}{60} = \dfrac{-17}{60} = -\dfrac{17}{60}$

APPLICATIONS

101.
 a. $\dfrac{5}{32} + \dfrac{1}{16} = \dfrac{5}{32} + \dfrac{2}{32} = \dfrac{7}{32}$ in

 b. $\dfrac{5}{32} - \dfrac{1}{16} = \dfrac{5}{32} - \dfrac{2}{32} = \dfrac{3}{32}$ in

103. $\dfrac{3}{8} + \dfrac{5}{16} = \dfrac{6}{16} + \dfrac{5}{16} = \dfrac{11}{16}$ in

105.
 a. $\dfrac{3}{8}$ not eaten.

 b. $\dfrac{2}{6} = \dfrac{1}{3}$ not eaten.

 c. $\dfrac{3}{8} + \dfrac{1}{3} = \dfrac{9}{24} + \dfrac{8}{24} = \dfrac{17}{24}$ not eaten.

 d. No – they ate more than a whole pizza.

107. The scale shows that it is undercharging by $\dfrac{1}{16}$ lb

109. $\dfrac{2}{5} + \dfrac{3}{10} = \dfrac{4}{10} + \dfrac{3}{10} = \dfrac{7}{10}$

111. $\dfrac{1}{8} + \dfrac{1}{4} + \dfrac{1}{16} + \dfrac{1}{8} + \dfrac{1}{4} + \dfrac{1}{16}$

$= \dfrac{2}{16} + \dfrac{4}{16} + \dfrac{1}{16} + \dfrac{2}{16} + \dfrac{4}{16} + \dfrac{1}{16}$

$= \dfrac{14}{16} = \dfrac{7}{8}$: no

113.

16/64	20/64
14/64	21/64

 a. The right rear has the most tread.

 b. The left rear has the least tread.

WRITING

115. Because they do not have the same denominator

REVIEW

117.
 a. $\dfrac{1}{4} + \dfrac{1}{8} = \dfrac{1}{4} \cdot \dfrac{2}{2} + \dfrac{1}{8} = \dfrac{2}{8} + \dfrac{1}{8} = \dfrac{3}{8}$

 b. $\dfrac{1}{4} - \dfrac{1}{8} = \dfrac{1}{4} \cdot \dfrac{2}{2} - \dfrac{1}{8} = \dfrac{2}{8} - \dfrac{1}{8} = \dfrac{1}{8}$

c. $\dfrac{1}{4} \cdot \dfrac{1}{8} = \dfrac{1}{32}$

d. $\dfrac{1}{4} \div \dfrac{1}{8} = \dfrac{1}{4} \cdot \dfrac{8}{1} = \dfrac{8}{4} = 2$

Section 3.5: Multiplying and Dividing Mixed Numbers

VOCABULARY

1. A <u>mixed</u> number, such as $8\dfrac{4}{5}$, is the sum of a whole number and a proper fraction.

3. The numerator of an <u>improper</u> fraction is greater than or equal to its denominator.

CONCEPTS

5. a. $5\dfrac{1}{3}°$ b. $-6\dfrac{7}{8} in.$

7. To write a mixed number as an improper fraction:

   1. <u>Multiply</u> the denominator of the fraction by the whole-number part.

   2. <u>Add</u> the numerator of the fraction to the result from Step 1.

   3. Write the sum from Step 2 over the original <u>denominator</u>.

9. $-\dfrac{4}{5}, -\dfrac{2}{5}, \dfrac{1}{5}$

11. To multiply or divide mixed numbers, first change the mixed numbers to <u>improper</u> fractions. Then perform the multiplication or division of the fractions as usual.

13. $4\dfrac{1}{5} \cdot 2\dfrac{5}{7} = 7\dfrac{2}{35}$ is not reasonable:

    $4\dfrac{1}{5} \cdot 2\dfrac{5}{7} \approx 4 \cdot 3 = 12$

NOTATION

15. a. We read $5\dfrac{11}{16}$ as "five <u>and</u> eleven-<u>sixteenths</u>."

b. We read $-4\dfrac{2}{3}$ as "<u>negative</u> four and <u>two</u>-thirds."

17. $5\dfrac{1}{4} \cdot 1\dfrac{1}{7} = \dfrac{21}{4} \cdot \dfrac{8}{7}$

    $= \dfrac{21 \cdot 8}{4 \cdot 7}$

    $= \dfrac{\cancel{7} \cdot 3 \cdot \cancel{4} \cdot 2}{\cancel{4} \cdot \cancel{7}}$

    $= \dfrac{6}{1}$

    $= 6$

GUIDED PRACTICE

19. $\dfrac{19}{8}; 2\dfrac{3}{8}$

21. $\dfrac{34}{25}; 1\dfrac{9}{25}$

23. $6\dfrac{1}{2} = \dfrac{2 \cdot 6 + 1}{2}$

    $= \dfrac{12 + 1}{2} = \dfrac{13}{2}$

25. $20\dfrac{4}{5} = \dfrac{5 \cdot 20 + 4}{5}$

    $= \dfrac{100 + 4}{5} = \dfrac{104}{5}$

27. $-7\dfrac{5}{9} = -\dfrac{9 \cdot 7 + 5}{9}$

    $= -\dfrac{63 + 5}{9} = -\dfrac{68}{9}$

29. $-8\dfrac{2}{3} = -\dfrac{3 \cdot 8 + 2}{3}$

    $= -\dfrac{24 + 2}{3} = -\dfrac{26}{3}$

31. $\dfrac{13}{4} = 3\dfrac{1}{4}$

33. $\dfrac{28}{5} = 5\dfrac{3}{5}$

35. $\dfrac{42}{9} = 4\dfrac{6}{9} = 4\dfrac{2\cdot\cancel{3}}{3\cdot\cancel{3}} = 4\dfrac{2}{3}$

37. $\dfrac{84}{8} = 10\dfrac{4}{8} = 10\dfrac{\cancel{4}}{\cancel{4}\cdot 2} = 10\dfrac{1}{2}$

39. $\dfrac{52}{13} = 4$

41. $\dfrac{34}{17} = 2$

43. $-\dfrac{58}{7} = -8\dfrac{2}{7}$

45. $-\dfrac{20}{6} = -3\dfrac{2}{6} = -3\dfrac{\cancel{2}}{\cancel{2}\cdot 3} = -3\dfrac{1}{3}$

47.

<--•--•--•--•---
  5 4 3 2 1 0 1 2 3 4

49.

<--•--•--•--•---
  4 3 2 1 0 1 2 3 4 5

51. $3\dfrac{1}{2}\cdot 2\dfrac{1}{3} = \dfrac{7}{2}\cdot\dfrac{7}{3} = \dfrac{49}{6} = 8\dfrac{1}{6}$

53. $2\dfrac{2}{5}\left(3\dfrac{1}{12}\right) = \dfrac{\cancel{12}}{5}\cdot\dfrac{37}{\cancel{12}}$
$= \dfrac{37}{5} = 7\dfrac{2}{5}$

55. $6\dfrac{1}{2}\cdot 1\dfrac{3}{13} = \dfrac{\cancel{13}}{2}\cdot\dfrac{16}{\cancel{13}} = \dfrac{16}{2} = 8$

57. $-2\dfrac{1}{2}(4) = -\dfrac{5}{2}\cdot\dfrac{4}{1} = -\dfrac{5}{\cancel{2}}\cdot\dfrac{\cancel{2}\cdot 2}{1} = -10$

59. $-1\dfrac{13}{15}\div\left(-4\dfrac{1}{5}\right) = -\dfrac{28}{15}\div\left(-\dfrac{21}{5}\right)$
$= -\dfrac{28}{15}\left(-\dfrac{5}{21}\right) = -\dfrac{4\cdot\cancel{7}}{3\cdot\cancel{5}}\left(-\dfrac{\cancel{5}}{3\cdot\cancel{7}}\right) = \dfrac{4}{9}$

61. $15\dfrac{1}{3}\div 2\dfrac{2}{9} = \dfrac{46}{3}\div\dfrac{20}{9} = \dfrac{46}{3}\cdot\dfrac{9}{20}$
$= \dfrac{\cancel{2}\cdot 23}{\cancel{3}}\cdot\dfrac{\cancel{3}\cdot 3}{\cancel{2}\cdot 10} = \dfrac{69}{10} = 6\dfrac{9}{10}$

63. $1\dfrac{3}{4}\div\dfrac{3}{4} = \dfrac{7}{4}\div\dfrac{3}{4} = \dfrac{7}{\cancel{4}}\cdot\dfrac{\cancel{4}}{3}$
$= \dfrac{7}{3} = 2\dfrac{1}{3}$

65. $1\dfrac{7}{24}\div\dfrac{7}{8} = \dfrac{31}{24}\div\dfrac{7}{8} = \dfrac{31}{24}\cdot\dfrac{8}{7}$
$= \dfrac{31}{3\cdot\cancel{8}}\cdot\dfrac{\cancel{8}}{7} = \dfrac{31}{21} = 1\dfrac{10}{21}$

TRY IT YOURSELF

67. $-6\cdot 2\dfrac{7}{24} = -\dfrac{6}{1}\cdot\dfrac{55}{24}$
$= -\dfrac{\cancel{6}}{1}\cdot\dfrac{55}{4\cdot\cancel{6}} = -\dfrac{55}{4} = -13\dfrac{3}{4}$

69. $-6\dfrac{3}{5}\div 7\dfrac{1}{3} = -\dfrac{33}{5}\div\dfrac{22}{3}$
$= -\dfrac{33}{5}\cdot\dfrac{3}{22} = -\dfrac{99}{110} = -\dfrac{9\cdot\cancel{11}}{10\cdot\cancel{11}} = -\dfrac{9}{10}$

71. $\left(1\dfrac{2}{3}\right)^2 = \left(\dfrac{5}{3}\right)^2 =$
$\left(\dfrac{5}{3}\right)\left(\dfrac{5}{3}\right) = \dfrac{25}{9} = 2\dfrac{7}{9}$

73. $8 \div 3\dfrac{1}{5} = \dfrac{8}{1} \div \dfrac{16}{5} = \dfrac{8}{1} \cdot \dfrac{5}{16}$

$= \dfrac{\cancel{8}}{1} \cdot \dfrac{5}{2 \cdot \cancel{8}} = \dfrac{5}{2} = 2\dfrac{1}{2}$

75. $-20\dfrac{1}{4} \div \left(-1\dfrac{11}{16}\right) = -\dfrac{81}{4} \div \left(-\dfrac{27}{16}\right)$

$= -\dfrac{81}{4} \cdot \left(-\dfrac{16}{27}\right) = -\dfrac{3 \cdot \cancel{27}}{\cancel{4}} \cdot \left(-\dfrac{\cancel{4} \cdot 4}{\cancel{27}}\right)$

$= 12$

77. $3\dfrac{1}{16} \cdot 4\dfrac{4}{7} = \dfrac{49}{16} \cdot \dfrac{32}{7} = \dfrac{\cancel{7} \cdot 7}{\cancel{16}} \cdot \dfrac{2 \cdot \cancel{16}}{\cancel{7}} = 14$

79. $-4\dfrac{1}{2} \div 2\dfrac{1}{4} = -\dfrac{9}{2} \div \dfrac{9}{4}$

$= -\dfrac{9}{2} \cdot \dfrac{4}{9} = -\dfrac{\cancel{9}}{\cancel{2}} \cdot \dfrac{\cancel{2} \cdot 2}{\cancel{9}} = -2$

81. $2\dfrac{1}{2}\left(-3\dfrac{1}{3}\right) = \dfrac{5}{2}\left(-\dfrac{10}{3}\right)$

$= -\dfrac{50}{6} = -8\dfrac{2}{6} = -8\dfrac{2}{2 \cdot 3}$

$= -8\dfrac{\cancel{2}}{\cancel{2} \cdot 3} = -8\dfrac{1}{3}$

83. $2\dfrac{5}{8} \cdot \dfrac{5}{27} = \dfrac{21}{8} \cdot \dfrac{5}{27}$

$= \dfrac{\cancel{3} \cdot 7}{8} \cdot \dfrac{5}{\cancel{3} \cdot 9} = \dfrac{35}{72}$

85. $6\dfrac{1}{4} \div 20 = \dfrac{25}{4} \div \dfrac{20}{1}$

$= \dfrac{25}{4} \cdot \dfrac{1}{20} = \dfrac{\cancel{5} \cdot 5}{4} \cdot \dfrac{1}{4 \cdot \cancel{5}}$

$= \dfrac{5}{16}$

87. $1\dfrac{2}{3} \cdot 6 \cdot \left(-\dfrac{1}{8}\right) = \dfrac{5}{3} \cdot \dfrac{6}{1} \left(-\dfrac{1}{8}\right)$

$= \dfrac{5}{\cancel{3}} \cdot \dfrac{\cancel{2} \cdot \cancel{3}}{1}\left(-\dfrac{1}{\cancel{2} \cdot 4}\right)$

$= -\dfrac{5}{4} = -1\dfrac{1}{4}$

89. $\left(-1\dfrac{1}{3}\right)^3 = \left(-\dfrac{4}{3}\right)^3 = \left(-\dfrac{4}{3}\right)\left(-\dfrac{4}{3}\right)\left(-\dfrac{4}{3}\right)$

$= -\dfrac{64}{27} = -2\dfrac{10}{27}$

APPLICATIONS

91. a. $3\dfrac{2}{3}$

b. $3\dfrac{2}{3} = \dfrac{3 \cdot 3 + 2}{3} = \dfrac{9+2}{3} = \dfrac{11}{3}$

93. Forward $2\dfrac{1}{2}$ somersaults.

95. a. $2\dfrac{2}{3}$

b. $2\dfrac{2}{3} - \dfrac{12}{3} = \dfrac{8}{3} - \dfrac{12}{3} = \dfrac{-4}{3} = -1\dfrac{1}{3}$

97. Since $59 < 60\dfrac{3}{4} < 61$ she needs a size 14.

Since $24\dfrac{1}{4} < 24\dfrac{1}{2} < 24\dfrac{3}{4}$ she needs a slim cut.

99. $12\dfrac{1}{4} \cdot 6\dfrac{1}{4} = \dfrac{25}{4} \cdot \dfrac{49}{4} = \dfrac{1225}{16} = 76\dfrac{9}{16} in$

101. $\dfrac{1}{2} \cdot 8\dfrac{1}{4} \cdot 10\dfrac{1}{3} = \dfrac{1}{2} \cdot \dfrac{33}{4} \cdot \dfrac{31}{3}$

$= \dfrac{341}{8} = 42\dfrac{5}{8} in$

103. $3\dfrac{1}{5} \cdot 20 = \dfrac{16}{5} \cdot \dfrac{20}{1} = \dfrac{16}{\cancel{5}} \cdot \dfrac{4 \cdot \cancel{5}}{1} = 64$ calories.

105. $84 \cdot 4\dfrac{1}{4} = 84 \cdot \dfrac{17}{4} = 357¢ = \$3.57$

107. $13\dfrac{3}{4} \div 11 = \dfrac{55}{4} \cdot \dfrac{1}{11} = \dfrac{5}{4} = 1\dfrac{1}{4}$ cups.

109. $200 \div \dfrac{1}{3} = \dfrac{200}{1} \div \dfrac{1}{3}$ people.
$= \dfrac{200}{1} \cdot \dfrac{3}{1} = 600$

111. $1\dfrac{1}{16} \div \dfrac{1}{8} = \dfrac{17}{16} \div \dfrac{1}{8} = \dfrac{17}{16} \cdot \dfrac{8}{1}$
$= \dfrac{17}{2 \cdot \cancel{8}} \cdot \dfrac{\cancel{8}}{1} = \dfrac{17}{2} = 8\dfrac{1}{2}$ furlongs.

WRITING

113. $2\dfrac{3}{4}$ represents $2 + \dfrac{3}{4}$ whereas $2\left(\dfrac{3}{4}\right)$ represents $2 \times \dfrac{3}{4}$.

REVIEW

115. $\left.\begin{array}{l} 5 = 5 \\ 12 = 2 \cdot 2 \cdot 3 \\ 15 = 3 \cdot 5 \end{array}\right\}$
$LCM = 2 \cdot 2 \cdot 3 \cdot 5 = 60$

117. $\left.\begin{array}{l} 12 = 2 \cdot 2 \cdot 3 \\ 68 = 2 \cdot 2 \cdot 17 \\ 92 = 2 \cdot 2 \cdot 23 \end{array}\right\}$
$GCF = 2 \cdot 2 = 4$

Section 3.6: Adding and Subtracting Mixed Numbers

VOCABULARY

1. A <u>mixed</u> number, such as $1\dfrac{7}{8}$, contains a whole number part and a fractional part.

3. To add (or subtract) mixed numbers written in vertical form, we add (or subtract) the <u>fractions</u> separately and the <u>whole</u> numbers separately.

5. Consider the following problem:

$$\begin{array}{r} 36\dfrac{5}{7} \\ +42\dfrac{4}{7} \\ \hline 78\dfrac{9}{7} \end{array} = 78 + 1\dfrac{2}{7} = 79\dfrac{2}{7}$$

Since we don't want an improper fraction in the answer we write $\dfrac{9}{7}$ as $1\dfrac{2}{7}$, <u>carry</u> the 1 and add it to 78 to get 79.

CONCEPTS

7. a. $76, \dfrac{3}{4}$ b. $76\dfrac{3}{4} = 76 + \dfrac{3}{4}$

9. a. 12 b. 30 c. 18 d. 24

NOTATION

11.
$$\begin{array}{r} 6\dfrac{3}{5} = 6\dfrac{3}{5} \cdot \dfrac{7}{7} = 6\dfrac{21}{35} \\ +3\dfrac{2}{7} = +3\dfrac{2}{7} \cdot \dfrac{5}{5} = +3\dfrac{10}{35} \\ \hline 9\dfrac{31}{35} \end{array}$$

GUIDED PRACTICE

13. $1\dfrac{1}{4} + 2\dfrac{1}{3} = \dfrac{5}{4} + \dfrac{7}{3} = \dfrac{5}{4} \cdot \dfrac{3}{3} + \dfrac{7}{3} \cdot \dfrac{4}{4}$
$= \dfrac{15}{12} + \dfrac{28}{12} = \dfrac{43}{12} = 3\dfrac{7}{12}$

15. $2\dfrac{1}{3} + 4\dfrac{2}{5} = \dfrac{7}{3} + \dfrac{22}{5} = \dfrac{7}{3} \cdot \dfrac{5}{5} + \dfrac{22}{5} \cdot \dfrac{3}{3}$
$= \dfrac{35}{15} + \dfrac{66}{15} = \dfrac{101}{15} = 6\dfrac{11}{15}$

17. $-4\dfrac{1}{8}+1\dfrac{3}{4}=\dfrac{-33}{8}+\dfrac{7}{4}=\dfrac{-33}{8}+\dfrac{7}{4}\cdot\dfrac{2}{2}$

$=\dfrac{-33}{8}+\dfrac{14}{8}=\dfrac{-19}{8}=-2\dfrac{3}{8}$

19. $-6\dfrac{5}{6}+3\dfrac{2}{3}=\dfrac{-41}{6}+\dfrac{11}{3}=\dfrac{-41}{6}+\dfrac{11}{3}\cdot\dfrac{2}{2}$

$=\dfrac{-41}{6}+\dfrac{22}{6}=\dfrac{-19}{6}=-3\dfrac{1}{6}$

21. $334\dfrac{1}{7}+42\dfrac{2}{3}=334+42+\dfrac{1}{7}+\dfrac{2}{3}$

$=376+\dfrac{1}{7}\cdot\dfrac{3}{3}+\dfrac{2}{3}\cdot\dfrac{7}{7}=376+\dfrac{3}{21}+\dfrac{14}{21}$

$=376\dfrac{17}{21}$

23. $667\dfrac{1}{5}+47\dfrac{3}{4}=667+47+\dfrac{1}{5}+\dfrac{3}{4}$

$=714+\dfrac{1}{5}\cdot\dfrac{4}{4}+\dfrac{3}{4}\cdot\dfrac{5}{5}=714+\dfrac{4}{20}+\dfrac{15}{20}$

$=714\dfrac{19}{20}$

25. $41\dfrac{2}{9}+18\dfrac{2}{5}=41+18+\dfrac{2}{9}+\dfrac{2}{5}$

$=59+\dfrac{2}{9}\cdot\dfrac{5}{5}+\dfrac{2}{5}\cdot\dfrac{9}{9}=59+\dfrac{10}{45}+\dfrac{18}{45}$

$=59\dfrac{28}{45}$

27. $89\dfrac{6}{11}+43\dfrac{1}{3}=89+43+\dfrac{6}{11}+\dfrac{1}{3}$

$=132+\dfrac{6}{11}\cdot\dfrac{3}{3}+\dfrac{1}{3}\cdot\dfrac{11}{11}=132+\dfrac{18}{33}+\dfrac{11}{33}$

$=132\dfrac{29}{33}$

29. $14\dfrac{1}{4}+29\dfrac{1}{20}+78\dfrac{3}{5}=14+29+78+\dfrac{1}{4}+\dfrac{1}{20}+\dfrac{3}{5}$

$=121+\dfrac{1}{4}\cdot\dfrac{5}{5}+\dfrac{1}{20}+\dfrac{3}{5}\cdot\dfrac{4}{4}=121+\dfrac{5}{20}+\dfrac{1}{20}+\dfrac{12}{20}$

$=121\dfrac{18}{20}=121\dfrac{9}{10}$

31. $106\dfrac{5}{18}+22\dfrac{1}{2}+19\dfrac{1}{9}=106+22+19+\dfrac{5}{18}+\dfrac{1}{2}$

$=147+\dfrac{5}{18}+\dfrac{1}{2}\cdot\dfrac{9}{9}+\dfrac{1}{9}\cdot\dfrac{2}{2}=147+\dfrac{5}{18}+\dfrac{9}{18}+\dfrac{2}{18}$

$=147\dfrac{16}{18}=147\dfrac{8}{9}$

33. $\begin{aligned}39\dfrac{5}{8}&=&39\dfrac{5}{8}\cdot\dfrac{3}{3}&=&39\dfrac{15}{24}\\+62\dfrac{11}{12}&=&+62\dfrac{11}{12}\cdot\dfrac{2}{2}&=&+62\dfrac{22}{24}\end{aligned}$

$101\dfrac{37}{24}=101+1\dfrac{13}{24}$

$=102\dfrac{13}{24}$

35. $\begin{aligned}82\dfrac{8}{9}&=&82\dfrac{8}{9}\cdot\dfrac{5}{5}&=&82\dfrac{40}{45}\\+46\dfrac{11}{15}&=&+46\dfrac{11}{15}\cdot\dfrac{3}{3}&=&+46\dfrac{33}{45}\end{aligned}$

$128\dfrac{73}{45}=128+1\dfrac{28}{45}$

$=129\dfrac{29}{45}$

37. $19\dfrac{11}{12}-9\dfrac{2}{3}=19\dfrac{11}{12}-9\dfrac{8}{12}=10\dfrac{3}{12}=10\dfrac{1}{4}$

39. $21\dfrac{5}{6}-8\dfrac{3}{10}=21\dfrac{25}{30}-8\dfrac{9}{30}=13\dfrac{16}{30}=13\dfrac{8}{15}$

41. $47\dfrac{1}{11}-15\dfrac{2}{3}=47\dfrac{3}{33}-15\dfrac{22}{33}$

$=46\dfrac{36}{33}-15\dfrac{22}{33}=31\dfrac{14}{33}$

43. $84\dfrac{5}{8}-12\dfrac{6}{7}=84\dfrac{35}{56}-12\dfrac{48}{56}$

$=83\dfrac{91}{56}-12\dfrac{48}{56}=71\dfrac{43}{56}$

45.
$$674 - 94\frac{11}{15} = 673\frac{15}{15} - 94\frac{11}{15}$$
$$= 579\frac{4}{15}$$

47.
$$112 - 49\frac{9}{32} = 111\frac{32}{32} - 49\frac{9}{32}$$
$$= 62\frac{23}{32}$$

TRY IT YOURSELF

49.
$$140\frac{5}{6} - 129\frac{4}{5} = 140\frac{25}{30} - 129\frac{24}{30} = 11\frac{1}{30}$$

51.
$$4\frac{1}{6} + 1\frac{1}{5} = \frac{25}{6} + \frac{6}{5} = \frac{125}{30} + \frac{36}{30}$$
$$= \frac{161}{30} = 5\frac{11}{30}$$

53.
$$5\frac{1}{2} + 3\frac{4}{5} = \frac{11}{2} + \frac{19}{5}$$
$$= \frac{55}{10} + \frac{38}{10} = \frac{93}{10} = 9\frac{3}{10}$$

55.
$$2 + 1\frac{7}{8} = \frac{16}{8} + \frac{15}{8} = \frac{31}{8} = 3\frac{7}{8}$$

57.
$$8\frac{7}{9} - 3\frac{1}{9} = 5\frac{6}{9} = 5\frac{2}{3}$$

59.
$$140\frac{3}{16} - 129\frac{3}{4}$$
$$= 139\frac{19}{16} - 129\frac{12}{16} = 10\frac{7}{16}$$

61.
$$380\frac{1}{6} + 17\frac{1}{4} = 380\frac{4}{24} + 17\frac{6}{24}$$
$$= 397\frac{10}{24} = 397\frac{5}{12}$$

63.
$$-2\frac{5}{6} + 1\frac{3}{8} = -\frac{17}{6} + \frac{11}{8}$$
$$= \frac{-136}{48} + \frac{66}{48} = -\frac{70}{48}$$
$$= -1\frac{22}{48} = -1\frac{11}{24}$$

65.
$$3\frac{1}{4} + 4\frac{1}{4} = 7\frac{2}{4} = 7\frac{1}{2}$$

67.
$$-3\frac{3}{4} + \left(-1\frac{1}{2}\right) = -\frac{15}{4} - \frac{3}{2}$$
$$= -\frac{15}{4} - \frac{6}{4} = -\frac{21}{4} = -5\frac{1}{4}$$

69.
$$7 - \frac{2}{3} = \frac{21}{3} - \frac{2}{3} = \frac{19}{3} = 6\frac{1}{3}$$

71.
$$12\frac{1}{2} + 5\frac{3}{4} + 35\frac{1}{6}$$
$$= 12\frac{6}{12} + 5\frac{9}{12} + 35\frac{2}{12}$$
$$= 52\frac{17}{12}$$
$$= 53\frac{5}{12}$$

73.
$$16\frac{1}{4} - 13\frac{3}{4} = 15\frac{5}{4} - 13\frac{3}{4} = 2\frac{2}{4} = 2\frac{1}{2}$$

75.
$$-4\frac{5}{8} - 1\frac{1}{4} = -\frac{37}{8} - \frac{5}{4}$$
$$= -\frac{37}{8} - \frac{10}{8} = -\frac{47}{8} = -5\frac{7}{8}$$

77.
$$6\frac{5}{8} - 3 = \frac{53}{8} - \frac{24}{8} = \frac{29}{8} = 3\frac{5}{8}$$

79.
$$\frac{7}{3} + 2 = \frac{7}{3} + \frac{6}{3} = \frac{13}{3} = 4\frac{1}{3}$$

81.
$$58\frac{7}{8} + 340\frac{1}{2} + 61\frac{3}{4}$$
$$= 58\frac{7}{8} + 340\frac{4}{8} + 61\frac{6}{8}$$
$$= 459\frac{17}{8}$$
$$= 461\frac{1}{8}$$

83. $9 - 8\frac{3}{4} = \frac{36}{4} - \frac{35}{4} = \frac{1}{4}$

APPLICATIONS

85. $3\frac{3}{4} + 1\frac{1}{2} = 3\frac{3}{4} + 1\frac{2}{4} = 4\frac{5}{4} = 5\frac{1}{4}\ hrs$

87.
$$2\frac{3}{4} + \frac{1}{2} + \frac{2}{3} + \frac{1}{3} + 2\frac{2}{3} + \frac{1}{4}$$
$$= \frac{11}{4} + \frac{1}{2} + \frac{2}{3} + \frac{1}{3} + \frac{8}{3} + \frac{1}{4}$$
$$= \frac{33}{12} + \frac{6}{12} + \frac{8}{12} + \frac{4}{12} + \frac{32}{12} + \frac{3}{12}$$
$$= \frac{86}{12} = 7\frac{2}{12} = 7\frac{1}{6}\ cups.$$

89.
$$2\frac{11}{16} + 2\frac{3}{4} + 3\frac{1}{4} + 2\frac{1}{2}$$
$$+ 1\frac{1}{2} + 2\frac{3}{4} + 1\frac{15}{16} + 2\frac{11}{16}$$
$$= 2\frac{11}{16} + 2\frac{12}{16} + 3\frac{4}{16} + 2\frac{8}{16}$$
$$+ 1\frac{8}{16} + 2\frac{12}{16} + 1\frac{15}{16} + 2\frac{11}{16}$$
$$= 15 + \frac{81}{16} = 15 + 5\frac{1}{16} = 20\frac{1}{16}\ lbs$$

91. $P = 2l + 2w$
$$P = 2\left(24\frac{1}{2}\right) + 2\left(29\frac{3}{4}\right)$$
$$P = \frac{98}{2} + \frac{119}{2} = \frac{217}{2}$$
$$P = 108\frac{1}{2}\ in.$$

93.
$$3\frac{1}{2} - \frac{3}{4} = \frac{7}{2} - \frac{3}{4}$$
$$= \frac{14}{4} - \frac{3}{4} = \frac{11}{4} = 2\frac{3}{4}\ mi$$

95. $50 - 1\frac{1}{2} = \frac{100}{2} - \frac{3}{2} = \frac{97}{2} = 48\frac{1}{2}\ ft.$

97. a. 20¢ b. 20¢

99.
$$311\frac{5}{12} - 119\frac{3}{4}$$
$$= 311\frac{5}{12} - 119\frac{9}{12}$$
$$= 310\frac{17}{12} - 119\frac{9}{12}$$
$$= 191\frac{8}{12} = 191\frac{2}{3}\ ft$$

WRITING

101. Opinion

103. Change $\frac{7}{5}$ into $1\frac{2}{5}$ and add the whole number parts to get $13\frac{2}{5}$.

REVIEW

105.
a. $3\frac{1}{2} + 1\frac{1}{4} = 3\frac{2}{4} + 1\frac{1}{4} = 4\frac{3}{4}$

b. $3\frac{1}{2} - 1\frac{1}{4} = 3\frac{2}{4} - 1\frac{1}{4} = 2\frac{1}{4}$

c. $3\frac{1}{2} \cdot 1\frac{1}{4} = \frac{7}{2} \cdot \frac{5}{4} = \frac{35}{8} = 4\frac{3}{8}$

d. $3\frac{1}{2} \div 1\frac{1}{4} = \frac{7}{2} \div \frac{5}{4} = \frac{7}{2} \cdot \frac{4}{5}$
$$= \frac{28}{10} = \frac{14}{5} = 2\frac{4}{5}$$

Section 3.7: Order of Operations and Complex Fractions

VOCABULARY

1. We use the order of <u>operations</u> rule to evaluate expressions that contain more than one operation.

3. $\dfrac{\frac{1}{2}}{\frac{2}{3}}$ and $\dfrac{\frac{7}{8}+\frac{2}{5}}{\frac{1}{2}-\frac{1}{3}}$ are examples of <u>complex</u> fractions.

CONCEPTS

5. Raising to a power, multiplication, and addition

7. $\left(\dfrac{2}{3}-\dfrac{1}{10}\right)+1\dfrac{2}{15}$ or $1\dfrac{2}{15}+\left(\dfrac{2}{3}-\dfrac{1}{10}\right)$

9. $\dfrac{2}{3} \div \dfrac{1}{5}$

11. $5\dfrac{3}{4} = \dfrac{23}{4}$

NOTATION

13. $\dfrac{7}{12}-\dfrac{1}{2}\cdot\dfrac{1}{3} = \dfrac{7}{12}-\dfrac{1\cdot 1}{2\cdot 3}$

$= \dfrac{7}{12}-\dfrac{1}{6}$

$= \dfrac{7}{12}-\dfrac{1}{6}\cdot\dfrac{2}{2}$

$= \dfrac{7}{12}-\dfrac{2}{12}$

$= \dfrac{5}{12}$

GUIDED PRACTICE

15. $\dfrac{3}{4}+\dfrac{2}{5}\left(-\dfrac{1}{2}\right)^2$

$= \dfrac{3}{4}+\dfrac{2}{5}\left(\dfrac{1}{4}\right)$

$= \dfrac{3}{4}+\dfrac{1}{10}$

$= \dfrac{15}{20}+\dfrac{2}{20}$

$= \dfrac{17}{20}$

17. $\dfrac{1}{6}+\dfrac{9}{8}\left(-\dfrac{2}{3}\right)^3$

$= \dfrac{1}{6}+\dfrac{9}{8}\left(-\dfrac{8}{27}\right)$

$= \dfrac{1}{6}-\dfrac{1}{3}$

$= \dfrac{1}{6}-\dfrac{2}{6}$

$= -\dfrac{1}{6}$

19. $\left(\dfrac{3}{4}-\dfrac{1}{6}\right) \div \left(-2\dfrac{1}{6}\right)$

$= \left(\dfrac{9}{12}-\dfrac{2}{12}\right) \div \left(-\dfrac{13}{6}\right)$

$= \left(\dfrac{7}{12}\right)\cdot\left(-\dfrac{6}{13}\right)$

$= -\dfrac{7}{26}$

21. $\left(\dfrac{15}{16}-\dfrac{1}{8}\right) \div \left(-9\dfrac{3}{4}\right)$

$= \left(\dfrac{15}{16}-\dfrac{2}{16}\right) \div \left(-\dfrac{39}{4}\right)$

$= \left(\dfrac{13}{16}\right)\cdot\left(-\dfrac{4}{39}\right)$

$= -\dfrac{1}{12}$

23. $\left(\dfrac{5}{6} - \dfrac{2}{3}\right) + 5\dfrac{4}{15}$

$= \left(\dfrac{5}{6} - \dfrac{4}{6}\right) + \dfrac{79}{15}$

$= \dfrac{1}{6} + \dfrac{79}{15}$

$= \dfrac{5}{30} + \dfrac{158}{30}$

$= \dfrac{163}{30}$

$= 5\dfrac{13}{30}$

25. $\left(\dfrac{7}{9} - \dfrac{1}{2}\right) + 2\dfrac{7}{18}$

$= \left(\dfrac{14}{18} - \dfrac{9}{18}\right) + \dfrac{43}{18}$

$= \dfrac{5}{18} + \dfrac{43}{18}$

$= \dfrac{48}{18}$

$= 2\dfrac{12}{18}$

$= 2\dfrac{2}{3}$

27. $A = \dfrac{1}{2}h(a+b)$

$A = \dfrac{1}{2}\left(5\dfrac{1}{4}\right)\left(2\dfrac{1}{2} + 7\dfrac{1}{2}\right)$

$A = \dfrac{1}{2}\left(\dfrac{21}{4}\right)(10)$

$A = \dfrac{210}{8}$

$A = 26\dfrac{1}{4}$

29. $A = \dfrac{1}{2}h(a+b)$

$A = \dfrac{1}{2}\left(4\dfrac{1}{2}\right)\left(1\dfrac{1}{4} + 6\dfrac{3}{4}\right)$

$A = \dfrac{1}{2}\left(\dfrac{9}{2}\right)(8)$

$A = 18$

31. $\dfrac{\frac{1}{16}}{\frac{2}{5}} = \dfrac{1}{16} \cdot \dfrac{5}{2} = \dfrac{5}{32}$

33. $\dfrac{\frac{5}{8}}{\frac{3}{4}} = \dfrac{5}{8} \cdot \dfrac{4}{3} = \dfrac{20}{24} = \dfrac{5}{6}$

35. $\dfrac{-\frac{1}{4} + \frac{2}{3}}{\frac{5}{6} + \frac{2}{3}}$

$= \dfrac{-\frac{3}{12} + \frac{8}{12}}{\frac{5}{6} + \frac{4}{6}}$

$= \dfrac{\frac{5}{12}}{\frac{9}{6}}$

$= \dfrac{5}{12} \cdot \dfrac{6}{9}$

$= \dfrac{5}{18}$

37.
$$\frac{\frac{1}{3}-\frac{3}{4}}{\frac{1}{6}+\frac{2}{3}}$$

$$=\frac{\frac{4}{12}-\frac{9}{12}}{\frac{1}{6}+\frac{4}{6}}$$

$$=\frac{-\frac{5}{12}}{\frac{5}{6}}$$

$$=-\frac{5}{12}\cdot\frac{6}{5}$$

$$=-\frac{1}{2}$$

39.
$$\frac{5-\frac{5}{6}}{1\frac{1}{12}}$$

$$=\frac{\frac{25}{6}}{\frac{13}{12}}$$

$$=\frac{25}{6}\cdot\frac{12}{13}$$

$$=3\frac{11}{13}$$

41.
$$\frac{4-\frac{7}{8}}{3\frac{1}{4}}$$

$$=\frac{\frac{25}{8}}{\frac{13}{4}}$$

$$=\frac{25}{8}\cdot\frac{4}{13}$$

$$=\frac{25}{26}$$

TRY IT YOURSELF

43. $\frac{7}{8}-\left(\frac{4}{5}+1\frac{3}{4}\right)$

$$=\frac{7}{8}-\left(\frac{4}{5}+\frac{7}{4}\right)$$

$$=\frac{7}{8}-\frac{51}{20}$$

$$=-\frac{67}{40}$$

$$=-1\frac{27}{40}$$

45.
$$\frac{-\frac{14}{15}}{\frac{7}{10}}$$

$$=-\frac{14}{15}\cdot\frac{10}{7}$$

$$=-\frac{4}{3}$$

$$=-1\frac{1}{3}$$

47. $A=\frac{1}{2}bh$

$A=\frac{1}{2}(10)\left(7\frac{1}{5}\right)$

$A=36$

49. $\frac{2}{3}\left(-\frac{1}{4}\right)+\frac{1}{2}$

$$=-\frac{1}{6}+\frac{1}{2}$$

$$=-\frac{1}{6}+\frac{3}{6}$$

$$=\frac{1}{3}$$

51. $\dfrac{4}{5} - \left(-\dfrac{1}{3}\right)^2$

$= \dfrac{4}{5} - \dfrac{1}{9}$

$= \dfrac{31}{45}$

53. $\dfrac{\dfrac{3}{8} + \dfrac{1}{4}}{\dfrac{3}{8} - \dfrac{1}{4}}$

$= \dfrac{\dfrac{5}{8}}{\dfrac{1}{8}}$

$= \dfrac{5}{8} \cdot \dfrac{8}{1}$

$= 5$

55. $\left(5\dfrac{1}{6} - 3\dfrac{7}{8}\right) + 12\dfrac{11}{12}$

$= \left(\dfrac{31}{6} - \dfrac{31}{8}\right) + \dfrac{155}{12}$

$= \dfrac{31}{24} + \dfrac{155}{12}$

$= \dfrac{341}{24}$

$= 14\dfrac{5}{24}$

57. $\dfrac{5\dfrac{1}{2}}{-\dfrac{1}{4} + \dfrac{3}{4}}$

$= \dfrac{\dfrac{11}{2}}{\dfrac{1}{2}}$

$= \dfrac{11}{2} \cdot \dfrac{2}{1}$

$= 11$

59. $\left|\dfrac{2}{3} - \dfrac{9}{10}\right| \div \left(-\dfrac{1}{5}\right)$

$= \left|-\dfrac{7}{30}\right| \cdot \left(-\dfrac{5}{1}\right)$

$= \dfrac{7}{30} \cdot (-5)$

$= -\dfrac{7}{6}$

$= -1\dfrac{1}{6}$

61. $\dfrac{\dfrac{1}{5} - \left(-\dfrac{1}{4}\right)}{\dfrac{1}{4} + \dfrac{4}{5}}$

$= \dfrac{\dfrac{4}{20} + \dfrac{5}{20}}{\dfrac{5}{20} + \dfrac{16}{20}}$

$= \dfrac{\dfrac{9}{20}}{\dfrac{21}{20}}$

$= \dfrac{9}{20} \cdot \dfrac{20}{21}$

$= \dfrac{3}{7}$

63. $1\dfrac{3}{5}\left(\dfrac{1}{2}\right)^2\left(\dfrac{3}{4}\right)$

$= \dfrac{8}{5} \cdot \dfrac{1}{4} \cdot \dfrac{3}{4}$

$= \dfrac{3}{10}$

65. $A = l \cdot w$

$A = 5\dfrac{5}{6} \cdot 7\dfrac{3}{5}$

$A = \dfrac{35}{6} \cdot \dfrac{38}{5}$

$A = \dfrac{133}{3}$

$A = 44\dfrac{1}{3}$

67.
$$\left(2-\frac{1}{2}\right)^2+\left(2+\frac{1}{2}\right)^2$$
$$=\left(\frac{3}{2}\right)^2+\left(\frac{5}{2}\right)^2$$
$$=\frac{9}{4}+\frac{25}{4}$$
$$=\frac{34}{4}$$
$$=8\frac{1}{2}$$

69.
$$\frac{-\frac{5}{6}}{-1\frac{7}{8}}$$
$$=\frac{-\frac{5}{6}}{-\frac{15}{8}}$$
$$=\left(-\frac{5}{6}\right)\cdot\left(-\frac{8}{15}\right)$$
$$=\frac{4}{9}$$

71.
$$\left(7\frac{3}{7}+3\frac{1}{5}\right)-9\frac{1}{10}$$
$$=\left(\frac{52}{7}+\frac{16}{5}\right)-\frac{91}{10}$$
$$=\left(\frac{372}{35}\right)-\frac{91}{10}$$
$$=\frac{744}{70}-\frac{637}{70}$$
$$=\frac{107}{70}$$
$$=1\frac{37}{70}$$

73.
$$\frac{\frac{1}{2}+\frac{1}{4}}{\frac{1}{2}-\frac{1}{4}}$$
$$=\frac{\frac{3}{4}}{\frac{1}{4}}$$
$$=\frac{3}{4}\cdot\frac{4}{1}$$
$$=3$$

75.
$$\left(\frac{8}{5}-1\frac{1}{3}\right)-\left(-\frac{4}{5}\cdot 10\right)$$
$$=\left(\frac{8}{5}-\frac{4}{3}\right)-(-8)$$
$$=\frac{4}{15}+8$$
$$=8\frac{4}{15}$$

APPLICATIONS

77.
$$20\left(4\frac{1}{2}+\frac{1}{16}\right)=20\left(4\frac{8}{16}+\frac{1}{16}\right)$$
$$=20\left(4\frac{9}{16}\right)=20\cdot\frac{73}{16}=5\cdot\frac{73}{4}$$
$$=\frac{365}{4}=91\frac{1}{4}in$$

79.
$$\frac{1}{16}+3\cdot\frac{1}{16}+\frac{5}{8}=\frac{1}{16}+\frac{3}{16}+\frac{10}{16}$$
$$=\frac{14}{16}=\frac{7}{8}oz$$

It can be mailed for the 1-ounce rate.

81.
$$5\left(\frac{1}{4}\right)+4\left(\frac{1}{2}\right)=\frac{5}{4}+2$$
$$=1\frac{1}{4}+2=3\frac{3}{4}hr$$

83. Total distance: $2\frac{4}{5}+1\frac{2}{5}+1\frac{4}{5}=4\frac{10}{5}=6mi$

$6 \div \frac{2}{3} = 6 \cdot \frac{3}{2} = \frac{18}{2} = 9$ parts.

85. $\frac{1}{2}+\frac{2}{3}+\frac{3}{4}=\frac{6}{12}+\frac{8}{12}+\frac{9}{12}=\frac{23}{12}$

$\frac{23}{12} \div \frac{1}{4} = \frac{23}{12} \cdot \frac{4}{1} = \frac{23}{3}$

7 used, $\frac{2}{3}$ tubes left over.

87. $A = \frac{1}{2}h(b_1 + b_2)$

$= \frac{1}{2} \cdot \left(2\frac{1}{3}\right)\left(2\frac{1}{2}+3\frac{1}{2}\right)$

$= \frac{1}{2} \cdot \frac{7}{3} \cdot 6 = 7 yd^2$

89. $\dfrac{1}{\frac{1}{10}+\frac{1}{15}} = \dfrac{1}{\frac{3}{30}+\frac{2}{30}}$

$= \dfrac{1}{\frac{5}{30}} = \frac{30}{5} = 6 \sec.$

WRITING

91. So that everyone will get the same answer if done correctly.

93. A fraction where the numerator and / or denominator has a fraction.

REVIEW

95. 2,248

97. $879 \cdot 23 = 20,217$

99. 1, 2, 3, 4, 6, 8, 12, 24

Chapter 3 Review

1. numerator 11, denominator 16, proper fraction

3. The figure is not divided into equal parts.

5. a. 1 b. 0
   c. 18 d. undefined

7. $\frac{2}{3} \cdot \frac{6}{6} = \frac{12}{18}$

9. $\frac{7}{15} \cdot \frac{3}{3} = \frac{21}{45}$

11. $5 = \frac{5}{1} \cdot \frac{9}{9} = \frac{45}{9}$

13. $\frac{15}{45} = \frac{1 \cdot \cancel{15}}{3 \cdot \cancel{15}} = \frac{1}{3}$

15. $\frac{66}{108} = \frac{\cancel{6} \cdot 11}{\cancel{6} \cdot 18} = \frac{11}{18}$

17. $\frac{81}{64}$ is in lowest terms – nothing divides 81 and 64.

19. Sleeping: $\frac{7}{24}$

    Not sleeping: $\frac{17}{24}$

21. To multiply two fractions, multiply the <u>numerators</u> and multiply the <u>denominators</u>. Then <u>simplify</u> if possible.

23. $\frac{1}{2} \cdot \frac{1}{3} = \frac{1 \cdot 1}{2 \cdot 3} = \frac{1}{6}$

25. $\frac{9}{16} \cdot \frac{20}{27} = \frac{\cancel{9} \cdot \cancel{4} \cdot 5}{4 \cdot \cancel{4} \cdot 3 \cdot \cancel{9}} = \frac{5}{4 \cdot 3} = \frac{5}{12}$

27. $\frac{3}{5} \cdot 7 = \frac{3}{5} \cdot \frac{7}{1} = \frac{21}{5}$

29. $3\left(\frac{1}{3}\right) = \frac{3}{1} \cdot \frac{1}{3} = \frac{3}{3} = 1$

31. $-\left(\frac{3}{4}\right)^2 = -\frac{3}{4} \cdot \frac{3}{4} = -\frac{9}{16}$

33. $\left(-\dfrac{2}{5}\right)^3 = \left(-\dfrac{2}{5}\right)\left(-\dfrac{2}{5}\right)\left(-\dfrac{2}{5}\right) = -\dfrac{8}{125}$

35. $8\left(\dfrac{1}{4}\right) = \dfrac{8}{1} \cdot \dfrac{1}{4} = \dfrac{8}{4} = 2\,mi$

37. $A = \dfrac{1}{2} \cdot b \cdot h$

    $A = \dfrac{1}{2} \cdot 15 \cdot 8$

    $A = \dfrac{1}{2} \cdot 120$

    $A = 60\,in^2$

39. a. 8

    b. $-\dfrac{12}{11}$

    c. $\dfrac{1}{5}$

    d. $\dfrac{7}{8}$

41. $\dfrac{1}{6} \div \dfrac{11}{25} = \dfrac{1}{6} \cdot \dfrac{25}{11} = \dfrac{25}{66}$

43. $-\dfrac{39}{25} \div \left(-\dfrac{13}{10}\right) = -\dfrac{39}{25}\left(-\dfrac{10}{13}\right)$

    $= \dfrac{3 \cdot \cancel{13} \cdot 2 \cdot \cancel{5}}{5 \cdot \cancel{5} \cdot \cancel{13}} = \dfrac{2 \cdot 3}{5} = \dfrac{6}{5}$

45. $-\dfrac{3}{8} \div \dfrac{1}{4} = -\dfrac{3}{8} \cdot \dfrac{4}{1} = -\dfrac{3 \cdot \cancel{4}}{2 \cdot \cancel{4}} = -\dfrac{3}{2}$

47. $\dfrac{2}{3} \div (-120) = -\dfrac{2}{3} \cdot \dfrac{1}{120} = -\dfrac{\cancel{2} \cdot 1}{3 \cdot \cancel{2} \cdot 60}$

    $= -\dfrac{1}{3 \cdot 60} = -\dfrac{1}{180}$

49. $\dfrac{3}{4} \div \dfrac{1}{16} = \dfrac{3}{4} \cdot \dfrac{16}{1} = \dfrac{3 \cdot \cancel{4} \cdot 4}{\cancel{4} \cdot 1} = 3 \cdot 4 = 12\,pins$

51. $\dfrac{2}{7} + \dfrac{3}{7} = \dfrac{2+3}{7} = \dfrac{5}{7}$

53. $\dfrac{7}{8} + \dfrac{3}{8} = \dfrac{7+3}{8} = \dfrac{10}{8} = \dfrac{\cancel{2} \cdot 5}{\cancel{2} \cdot 4} = \dfrac{5}{4}$

55. a. $\dfrac{3}{8} + \dfrac{2}{8} = \dfrac{3+2}{8} = \dfrac{5}{8}$

    b. $\dfrac{4}{5} - \dfrac{3}{5} = \dfrac{4-1}{5} = \dfrac{3}{5}$

57. $\dfrac{1}{6} + \dfrac{2}{3} = \dfrac{1}{6} + \dfrac{2}{3} \cdot \dfrac{2}{2} = \dfrac{1}{6} + \dfrac{4}{6} = \dfrac{5}{6}$

59. $\dfrac{5}{24} + \dfrac{3}{16} = \dfrac{5}{24} \cdot \dfrac{2}{2} + \dfrac{3}{16} \cdot \dfrac{3}{3}$

    $= \dfrac{10}{48} + \dfrac{9}{48} = \dfrac{19}{48}$

61. $-\dfrac{19}{18} + \dfrac{5}{12} = -\dfrac{19}{18} \cdot \dfrac{2}{2} + \dfrac{5}{12} \cdot \dfrac{3}{3}$

    $= -\dfrac{38}{36} + \dfrac{15}{36} = -\dfrac{23}{36}$

63. $-6 + \dfrac{13}{6} = -\dfrac{6}{1} \cdot \dfrac{6}{6} + \dfrac{13}{6}$

    $= -\dfrac{36}{6} + \dfrac{13}{6} = -\dfrac{23}{6}$

65. $\dfrac{3}{4} - \dfrac{17}{32} = \dfrac{3}{4} \cdot \dfrac{8}{8} - \dfrac{17}{32}$

    $= \dfrac{24}{32} - \dfrac{17}{32} = \dfrac{7}{32}\,in$

67. $\dfrac{2}{9}\,\square\,\dfrac{3}{11}$

    $\dfrac{2}{9} \cdot \dfrac{11}{11}\,\square\,\dfrac{3}{11} \cdot \dfrac{9}{9}$

    $\dfrac{22}{99} < \dfrac{27}{99}$

    The second hour was better.

69. $4\dfrac{1}{4} = \dfrac{4 \cdot 4 + 1}{4} = \dfrac{17}{4}$

71.

$$5\overline{)16}$$
$$\underline{-15}$$
$$1$$

$$\frac{16}{5} = 3\frac{1}{5}$$

73.

$$3\overline{)51}$$
$$\underline{-51}$$
$$0$$

$$\frac{51}{3} = 17$$

75. $9\frac{3}{8} = \frac{8 \cdot 9 + 3}{8} = \frac{72 + 3}{8} = \frac{75}{8}$

77. $3\frac{11}{14} = \frac{14 \cdot 3 + 11}{14} = \frac{42 + 11}{14} = \frac{53}{14}$

79. $1\frac{2}{5} \cdot 1\frac{1}{2} = \frac{7}{5} \cdot \frac{3}{2} = \frac{21}{10} = 2\frac{1}{10}$

81. $-6\left(-6\frac{2}{3}\right) = -\frac{6}{1}\left(-\frac{20}{3}\right)$

$= \frac{2 \cdot \cancel{3} \cdot 20}{1 \cdot \cancel{3}} = 2 \cdot 20 = 40$

83. $-11\frac{1}{5} \div \left(-\frac{7}{10}\right) = \left(-\frac{56}{5}\right) \cdot \left(-\frac{10}{7}\right)$

$= \frac{\cancel{7} \cdot 8 \cdot 2 \cdot \cancel{5}}{\cancel{5} \cdot \cancel{7}} = 8 \cdot 2 = 16$

85. $\left(-2\frac{3}{4}\right)^2 = \left(-\frac{11}{4}\right)^2 = \left(-\frac{11}{4}\right)\left(-\frac{11}{4}\right)$

$= \frac{121}{16} = 7\frac{9}{16}$

87. $5\frac{1}{2}\left(8\frac{3}{4}\right) = \frac{11}{2} \cdot \frac{35}{4} = \frac{385}{8}$

$= 48\frac{1}{8} in$

89. $90 \div 2\frac{1}{4} = 90 \div \frac{9}{4} = \frac{90}{1} \cdot \frac{4}{9}$

$= \frac{\cancel{9} \cdot 10 \cdot 4}{1 \cdot \cancel{9}} = 4 \cdot 10 = 40$ posters

91. $1\frac{3}{8} + 2\frac{1}{5} = \frac{11}{8} + \frac{11}{5} = \frac{11}{8} \cdot \frac{5}{5} + \frac{11}{5} \cdot \frac{8}{8}$

$= \frac{55}{40} + \frac{88}{40} = \frac{143}{40} = 3\frac{23}{40}$

93. $2\frac{5}{6} - 1\frac{3}{4} = \frac{17}{6} - \frac{7}{4} = \frac{17}{6} \cdot \frac{2}{2} - \frac{7}{4} \cdot \frac{3}{3}$

$= \frac{34}{12} - \frac{21}{12} = \frac{13}{12} = 1\frac{1}{12}$

95.

$$157\frac{11}{30} = 157\frac{11}{30} \cdot \frac{2}{2} = 157\frac{22}{60}$$
$$+98\frac{7}{12} \quad +98\frac{7}{12} \cdot \frac{5}{5} \quad +98\frac{35}{60}$$
$$\phantom{+98\frac{7}{12}} = 255\frac{57}{60}$$
$$\phantom{+98\frac{7}{12}} = 255\frac{19}{20}$$

97.

$$33\frac{8}{9} = 33\frac{8}{9} \cdot \frac{2}{2} = 33\frac{16}{18}$$
$$+49\frac{1}{6} \quad +49\frac{1}{6} \cdot \frac{3}{3} \quad +49\frac{3}{18}$$
$$\phantom{+49\frac{1}{6}} = 82\frac{19}{18}$$
$$\phantom{+49\frac{1}{6}} = 82 + 1\frac{1}{18}$$
$$\phantom{+49\frac{1}{6}} = 83\frac{1}{18}$$

99.

$$50\frac{5}{8} = 50\frac{5}{8} \cdot \frac{3}{3} = 50\frac{15}{24}$$
$$-19\frac{1}{6} \quad -19\frac{1}{6} \cdot \frac{4}{4} \quad -19\frac{4}{24}$$
$$\phantom{-19\frac{1}{6}} = 31\frac{11}{24}$$

101.
$$23\frac{1}{3} = 23\frac{1}{3} \cdot \frac{2}{2} = 23\frac{2}{6} = 22\frac{8}{6}$$
$$-2\frac{5}{6} \quad -2\frac{5}{6} \quad -2\frac{5}{6} \quad -2\frac{5}{6}$$
$$= 20\frac{3}{6}$$
$$= 20\frac{1}{2}$$

103.
$$10\frac{3}{4} + 21\frac{1}{2} + 7\frac{2}{3} = 10\frac{3}{4} \cdot \frac{3}{3} + 21\frac{1}{2} \cdot \frac{6}{6} + 7\frac{2}{3} \cdot \frac{2}{2}$$
$$= 10\frac{9}{12} + 21\frac{6}{12} + 7\frac{8}{12} = 38\frac{23}{12}$$
$$= 38 + 1\frac{11}{12} = 39\frac{11}{12} \text{ gal}$$

105.
$$\frac{3}{4} + \left(-\frac{1}{3}\right)^2 \left(\frac{5}{4}\right) = \frac{3}{4} + \left(\frac{1}{9}\right)\left(\frac{5}{4}\right)$$
$$= \frac{3}{4} + \frac{5}{36} = \frac{3}{4} \cdot \frac{9}{9} + \frac{5}{36} = \frac{27+5}{36}$$
$$= \frac{32}{36} = \frac{8}{9}$$

107.
$$\left(\frac{11}{5} - 1\frac{2}{3}\right) - \left(-\frac{4}{9} \cdot 18\right)$$
$$= \left(\frac{11}{5} - \frac{5}{3}\right) + \left(\frac{4}{9} \cdot \frac{18}{1}\right)$$
$$= \left(\frac{33}{15} - \frac{25}{15}\right) + \left(\frac{4}{\cancel{9}} \cdot \frac{2 \cdot \cancel{9}}{1}\right)$$
$$= \frac{8}{15} + 8 = 8\frac{8}{15}$$

109.
$$\frac{\frac{3}{5}}{-\frac{17}{20}} = \frac{3}{5}\left(-\frac{20}{17}\right) = \frac{3}{\cancel{5}}\left(-\frac{4 \cdot \cancel{5}}{17}\right) = -\frac{12}{17}$$

111.
$$\frac{\frac{2}{3} - \frac{1}{6}}{\frac{3}{4} - \frac{1}{2}} = \frac{\frac{2}{3} \cdot \frac{2}{2} - \frac{1}{6}}{\frac{3}{4} - \frac{1}{2} \cdot \frac{2}{2}} = \frac{\frac{4}{6} - \frac{1}{6}}{\frac{3}{4} - \frac{2}{4}}$$
$$= \frac{\frac{3}{6}}{\frac{5}{4}} = \frac{\frac{1}{2}}{\frac{5}{4}} = \frac{1}{2}\left(-\frac{4}{5}\right) = -\frac{\cancel{2} \cdot 2}{\cancel{2} \cdot 5}$$
$$= -\frac{2}{5}$$

113.
$$\left(5\frac{1}{5} + 1\frac{1}{2}\right) - 4\frac{1}{8} = \frac{26}{5} + \frac{3}{2} - \frac{33}{8}$$
$$= \frac{26}{5} \cdot \frac{8}{8} + \frac{3}{2} \cdot \frac{20}{20} - \frac{33}{8} \cdot \frac{5}{5} = \frac{208}{40} + \frac{60}{40} - \frac{165}{40}$$
$$= \frac{103}{40} = 2\frac{23}{40}$$

115.
$$A = \frac{1}{2}h(a+b) = \frac{1}{2} \cdot 2\frac{7}{9}\left(1\frac{1}{8} + 4\frac{7}{8}\right)$$
$$= \frac{1}{2} \cdot \frac{25}{9} \cdot 5\frac{8}{8} = \frac{1}{2} \cdot \frac{25}{9} \cdot 6 = \frac{25}{\underset{3}{\cancel{9}}} \cdot \cancel{6}$$
$$= \frac{25}{3} = 8\frac{1}{3}$$

117.
$$1\frac{1}{2} + 2\frac{2}{3} + \frac{3}{4} = \frac{3}{2} + \frac{8}{3} + \frac{3}{4}$$
$$= \frac{3}{2} \cdot \frac{6}{6} + \frac{8}{3} \cdot \frac{4}{4} + \frac{3}{4} \cdot \frac{3}{3} = \frac{18}{12} + \frac{32}{12} + \frac{9}{12}$$
$$= \frac{59}{12} \text{ oz cream}$$

$$\frac{59}{12} \div \frac{5}{6} = \frac{59}{\underset{2}{\cancel{12}}} \cdot \frac{\cancel{6}}{5} = \frac{59}{10} = 5\frac{9}{10}$$

He can make 5 full tubes with $\frac{9}{10}$ of a tube left over.

Chapter 3 Test

1. a. For the fraction $\dfrac{6}{7}$, the <u>numerator</u> is 6 and the <u>denominator</u> is 7.

    b. Two fractions are <u>equivalent</u> if they represent the same number.

    c. A fraction is in <u>simplest</u> form when the numerator and denominator have no common factors other than 1.

    d. To <u>simplify</u> a fraction, we remove common factors of the numerator and denominator.

    e. The <u>reciprocal</u> of $\dfrac{4}{5}$ is $\dfrac{5}{4}$.

    f. A <u>mixed</u> number, such as $1\dfrac{9}{16}$, is the sum of a whole number and a proper fraction.

    g. $\dfrac{\frac{1}{8}}{\frac{7}{12}}$ and $\dfrac{\frac{3}{4}+\frac{1}{3}}{\frac{5}{12}-\frac{1}{4}}$ are examples of <u>complex</u> fractions.

3. $2\dfrac{1}{6} = \dfrac{2\cdot 6+1}{6} = \dfrac{13}{6}$

5. Yes: $\dfrac{5}{15} = \dfrac{1\cdot \cancel{5}}{3\cdot \cancel{5}} = \dfrac{1}{3}$

7. a. 0   b. undefined

9. $\dfrac{3}{16} + \dfrac{7}{16} = \dfrac{10}{16} = \dfrac{\cancel{2}\cdot 5}{\cancel{2}\cdot 8} = \dfrac{5}{8}$

11. $\dfrac{4}{3} \div \dfrac{2}{9} = \dfrac{4}{3}\cdot\dfrac{9}{2} = \dfrac{\cancel{2}\cdot 2\cdot \cancel{3}\cdot 3}{\cancel{3}\cdot \cancel{2}} = 2\cdot 3 = 6$

13. $-\dfrac{3}{7}+2 = -\dfrac{3}{7}+\dfrac{2}{1}\cdot\dfrac{7}{7} = -\dfrac{3}{7}+\dfrac{14}{7} = \dfrac{11}{7}$

15. $\dfrac{8}{9}\square\dfrac{9}{10}$

    $\dfrac{8}{9}\cdot\dfrac{10}{10}\square\dfrac{9}{10}\cdot\dfrac{9}{9}$

    $\dfrac{80}{90} < \dfrac{81}{90}$

    so the larger fraction is $\dfrac{9}{10}$

17. $\dfrac{16}{25}+\dfrac{1}{5}+\dfrac{1}{10} = \dfrac{16}{25}\cdot\dfrac{2}{2}+\dfrac{1}{5}\cdot\dfrac{10}{10}+\dfrac{1}{10}\cdot\dfrac{5}{5}$

    $= \dfrac{32}{50}+\dfrac{10}{50}+\dfrac{5}{50} = \dfrac{47}{50}$

19. $\begin{aligned}157\dfrac{3}{10} &= 157\dfrac{3}{10}\cdot\dfrac{3}{3} = 157\dfrac{9}{30}\\ +103\dfrac{13}{15} &\quad +103\dfrac{13}{15}\cdot\dfrac{2}{2} \quad +103\dfrac{26}{30}\\ &\qquad\qquad\qquad\qquad\qquad = 260\dfrac{35}{30}\\ &\qquad\qquad\qquad\qquad\qquad = 260+1\dfrac{5}{30}\\ &\qquad\qquad\qquad\qquad\qquad = 261\dfrac{1}{6}\end{aligned}$

21. $6\dfrac{1}{4}\div 3\dfrac{3}{4} = \dfrac{25}{4}\div\dfrac{15}{4} = \dfrac{25}{4}\cdot\dfrac{4}{15}$

    $= \dfrac{\cancel{5}\cdot 5\cdot\cancel{4}}{\cancel{4}\cdot 3\cdot\cancel{5}} = \dfrac{5}{3} = 1\dfrac{2}{3}$

23. $P = 2l+2w = 2\cdot\dfrac{1}{3}+2\cdot\dfrac{1}{9} = \dfrac{2}{3}+\dfrac{2}{9}$

    $= \dfrac{2}{3}\cdot\dfrac{3}{3}+\dfrac{2}{9} = \dfrac{6}{9}+\dfrac{2}{9} = \dfrac{8}{9}$

25. $10\dfrac{1}{2}+2\left(\dfrac{5}{8}\right) = \dfrac{21}{2}+\dfrac{10}{8} = \dfrac{21}{2}\cdot\dfrac{4}{4}+\dfrac{10}{8}$

    $= \dfrac{84}{8}+\dfrac{10}{8} = \dfrac{94}{8} = 11\dfrac{6}{8} = 11\dfrac{3}{4}in$

27. $40\left(1\dfrac{1}{2}\right) = 40\left(\dfrac{3}{2}\right) = \dfrac{120}{2} = 60$ calories

29.
$$\left(\frac{2}{3}\cdot\frac{5}{16}\right)-\left(-1\frac{3}{5}\div 4\frac{4}{5}\right)$$
$$=\frac{10}{48}-\left(-\frac{8}{5}\div\frac{24}{5}\right)$$
$$=\frac{5}{24}+\left(\frac{8}{5}\cdot\frac{5}{24}\right)=\frac{5}{24}+\frac{40}{120}=\frac{5}{24}+\frac{1}{3}$$
$$=\frac{5}{24}+\frac{1}{3}\cdot\frac{8}{8}=\frac{5}{24}+\frac{8}{24}=\frac{13}{24}$$

31.
$$\frac{\frac{5}{6}}{\frac{7}{8}}=\frac{5}{6}\cdot\frac{8}{7}=\frac{5\cdot\cancel{2}\cdot 4}{\cancel{2}\cdot 3\cdot 7}=\frac{20}{21}$$

33. $\frac{3}{4}\cdot\frac{4}{3}=\frac{12}{12}=1$ This holds for every number and its reciprocal since we are multiplying the same two numbers in the numerator and denominator, just in different orders.

Chapters 1 – 3 Cumulative Review

1. a. 5

b. 8 hundred thousands

c. 5,896,600

d. 5,900,000

3. Orange, San Diego, Kings, Miami-Dade, Dallas, Queens

5.
```
  2 2 1 2
  7,897
  6,909
  1,812
+14,378
 30,996
```

7.
```
    20
   ×12
    40
  +200
   240
```

```
   240
   ×10
 2,400
```

2,400 stickers per packet

9.
```
       991
   35)34685
      -315
       318
      -315
        35
       -35
         0
```
$35\cdot 991 = 34,685$

11. 1, 2, 3, 4, 6, 8, 12, 24

13. $16 = 2\cdot 2\cdot 2\cdot 2$
$20 = 2\cdot 2\cdot 5$
$LCM(16, 20) = 2\cdot 2\cdot 2\cdot 2\cdot 5 = 80$

15.
$$15+5\left[12-(2^2+4)\right]$$
$$=15+5\left[12-(4+4)\right]$$
$$=15+5[12-8]$$
$$=15+5[4]$$
$$=15+20$$
$$=35$$

17. $\{...,-3,-2,-1,0,1,2,3,...\}$

19. $-20+6+(-1) = -14+(-1) = -15$

21. $-25(8) = -200$ ft

23. $6+(-2)(-5)=6+10=16$

25. $-5+3|-4-(-6)|$
$=-5+3|-4+6|$
$=-5+3|2|$
$=-5+6$
$=1$

27. $\dfrac{21}{28}=\dfrac{3\cdot\cancel{7}}{4\cdot\cancel{7}}=\dfrac{3}{4}$

29. $\dfrac{6}{5}\left(-\dfrac{2}{3}\right)=-\dfrac{2\cdot\cancel{3}\cdot2}{5\cdot\cancel{3}}=-\dfrac{2\cdot2}{5}=-\dfrac{4}{5}$

31. $\dfrac{2}{3}+\dfrac{3}{4}=\dfrac{2}{3}\cdot\dfrac{4}{4}+\dfrac{3}{4}\cdot\dfrac{3}{3}$
$=\dfrac{8}{12}+\dfrac{9}{12}=\dfrac{17}{12}=1\dfrac{5}{12}$

33. $\dfrac{1}{3}(90)=\dfrac{1}{\cancel{3}}\cdot\dfrac{\cancel{3}\cdot30}{1}=\dfrac{30}{1}=30\,\text{sec}$

35.
$$\begin{array}{r}10\\7\overline{)75}\\-70\\\hline 5\end{array}$$

$\dfrac{75}{7}=10\dfrac{5}{7}$

37. $2\dfrac{2}{5}\left(3\dfrac{1}{12}\right)=\dfrac{\cancel{12}}{5}\cdot\dfrac{37}{\cancel{12}}=\dfrac{37}{5}=7\dfrac{2}{5}$

39. $4\dfrac{2}{3}+5\dfrac{1}{4}=4\dfrac{2}{3}\cdot\dfrac{4}{4}+5\dfrac{1}{4}\cdot\dfrac{3}{3}$
$=4\dfrac{8}{12}+5\dfrac{3}{12}=9\dfrac{11}{12}$

41. $4\left(1\dfrac{1}{2}\right)$ by $8\left(3\dfrac{1}{2}\right)$
$=4\cdot\dfrac{3}{2}$ by $8\cdot\dfrac{7}{2}$
$=6\,in$ by $28\,in$

43. $1\dfrac{1}{3}+1\dfrac{1}{3}+\dfrac{3}{4}=2\dfrac{2}{3}+\dfrac{3}{4}$
$=2\dfrac{2}{3}\cdot\dfrac{4}{4}+\dfrac{3}{4}\cdot\dfrac{3}{3}$
$=2\dfrac{8}{12}+\dfrac{9}{12}=2\dfrac{17}{12}$
$=2+1\dfrac{5}{12}=3\dfrac{5}{12}\,ft$

45. $\dfrac{\dfrac{2}{3}}{\dfrac{4}{5}}=\dfrac{2}{3}\cdot\dfrac{5}{4}=\dfrac{10}{12}=\dfrac{5}{6}$

CHAPTER 4: Decimals

Section 4.1: An Introduction to Decimals

VOCABULARY

1. Decimals are written by entering the digits 0, 1, 2, 3, 4, 5, 6, 7, 8, and 9 into place-value columns that are separated by a decimal <u>point</u>.

3. We can show the value represented by each digit of the decimal 98.6213 by using <u>expanded</u> form:

$$90 + 8 + \frac{6}{10} + \frac{2}{100} + \frac{1}{1,000} + \frac{3}{10,000}$$

CONCEPTS

5. *From left to right*: Thousands, Hundreds, Tens, Ones, Tenths, Hundredths, Thousandths, Ten-thousandths.

7. a. The value of each place in the whole-number part of a decimal number is <u>10</u> times greater than the column directly to its right.

   b. The value of each place in the fractional part of a decimal number is $\frac{1}{10}$ of the value of the place directly to its left.

9. a. $\frac{7}{10}$; 0.7

   b. $\frac{47}{100}$; 0.47

11. Whole-number part ; Fractional part

NOTATION

13. The columns to the right of the decimal point in a decimal number form its fractional part. Their place value names are similar to those in the whole-number part, but they end in the letters "<u>ths</u>."

15. 79,816.0245

GUIDED PRACTICE

17. a. 9 tenths

    b. 6

    c. 4

    d. 5 ones

19. a. 8 millionths

    b. 0

    c. 5

    d. 6 ones

21. $37.89 = 30 + 7 + \frac{8}{10} + \frac{9}{100}$

23. $124.575 = 100 + 20 + 4 + \frac{5}{10} + \frac{7}{100} + \frac{5}{1,000}$

25. $7,498.6468 = 7000 + 400 + 90 + 8 + \frac{6}{10} + \frac{4}{100} + \frac{6}{1,000} + \frac{8}{10,000}$

27. $6.40941 = 6 + \frac{4}{10} + \frac{0}{100} + \frac{9}{1,000} + \frac{4}{10,000} + \frac{1}{100,000}$

29. $0.3 = \text{three tenths} = \frac{3}{10}$

31. $50.41 = \text{fifty and forty-one hundredths} = 50\frac{41}{100}$

33. $19.529 = \text{nineteen and five hundred twenty-nine thousandths} = 19\frac{529}{1,000}$

35. $304.0003 = \text{three hundred four and three ten-thousandths} = 304\frac{3}{10,000}$

37. -0.00137 = negative one hundred thirty-seven hundred-thousandths = $-\dfrac{137}{100,000}$

39. $-1,072.499$ = negative one thousand seventy-two and four hundred ninety-nine thousandths = $-1,072\dfrac{499}{1,000}$

41. 6.187

43. 10.0056

45. -16.39

47. 104.000004

49. $2.59 > 2.55$

51. $45.103 < 45.108$

53. $3.28724 > 3.2871$

55. $379.67 > 379.6088$

57. $-23.45 < -23.1$

59. $-0.065 > -0.066$

61.

63.

65. 506.2

67. 33.08

69. 4.234

71. 0.3656

73. -0.14

75. -2.7

77. 3.150

79. 1.414213

81. 16.100

83. 290.30350

85. $0.28

87. $27,841

APPLICATIONS

89. -0.7

91. The blank is $1,025.78.

93. ---------|---------|---------|-------*-|---------|
       .1   .2   .3   .4   .5

95. two-thousandths, $\dfrac{2}{1,000} = \dfrac{1}{500}$

97. $0.16, $1.02, $1.20, $0.00, $0.10

99. Candlemaking, Crafts, Hobbies, Folk Dolls, Modern Art

101. Cylinders 2 and 4 should be replaced.

103. bacterium, plant cell, animal cell, asbestos fiber

105. a. Quarter 3 of 2007: a gain of $2.75

    b. Quarter 4 of 2006: a loss of $2.05

WRITING

107. Ten is ten times larger than one, while one-tenth is ten times smaller.

109. Because the phrase would be translated as $102\dfrac{3}{1,000}$.

111. decade : 10 years

decathlon : 10 events

decimal : 10 parts

REVIEW

113. a. Perimeter

$$3\frac{1}{2}+2\frac{3}{4}+3\frac{1}{2}+2\frac{3}{4}$$
$$=\frac{7}{2}+\frac{11}{4}+\frac{7}{2}+\frac{11}{4}$$
$$=\frac{14}{4}+\frac{11}{4}+\frac{14}{4}+\frac{11}{4}$$
$$=\frac{50}{4}$$
$$=12\frac{1}{2}\,ft.$$

b. Area

$$3\frac{1}{2}\cdot 2\frac{3}{4}=\frac{7}{2}\cdot\frac{11}{4}=\frac{77}{8}=9\frac{5}{8}\,ft^2$$

Section 4.2: Adding and Subtracting Decimals

VOCABULARY

1. 1.72 ← addend
4.68 ← addend
+2.02 ← addend
8.42 ← sum

3. 12.9 ← minuend
−4.3 ← subtrahend
8.6 ← difference

5. To see whether the result of an addition is reasonable, we can round the addends and estimate the sum or difference.

CONCEPTS

7. The subtraction is incorrect: $15.2+12.5 \ne 28.7$

9. To subtract signed decimals, add the opposite of the decimal that is being subtracted.

11. a. $6.8-1.2=6.8+(-1.2)$

b. $29.03-(-13.55)=29.03+13.55$

c. $-5.1-7.4=-5.1+(-7.4)$

NOTATION

13. 46.600
11.000
+15.702

GUIDED PRACTICE

15. 32.5
+7.4
39.9

17. 3.04
4.12
+1.43
8.59

19. $\overset{2\ 1}{3}6.821$
7.300
42.000
+15.440
101.561

21. $\overset{1\ 2}{2}7.471$
6.400
157.000
+ 12.120
202.991

23. 6.83
−3.52
3.31

25.   8.97
     −6.22

      2.75

27.   ⁴ ¹⁴
      49̶5̶.4
     −153.7

      341.7

29.   ⁷ ¹¹
      87̶8̶.1
     −174.6

      703.5

31.   ⁹
       2̶ 1̶0 10
      18.3̶0̶0̶
     −11.065

       7.235

33.   ⁹
        8 1̶0 10
      66.9̶0̶0̶
     −23.037

      43.863

35. $-6.3+(-8.4)=-14.7$

37. $-9.5+(-9.3)=-18.8$

39. $4.12+(-18.8)=-14.68$

41. $6.45+(-12.6)=-6.15$

43. $-62.8-3.9=-62.8+(-3.9)=-66.7$

45. $-42.5-2.8=-42.5+(-2.8)=-45.3$

47. $-4.49-(-11.3)=-4.49+11.3=6.81$

49. $-6.78-(-24.6)=-6.78+24.6=17.82$

51. $-11.1-(-14.4+7.8)=-11.1-(-6.6)$
    $=-11.1+6.6=-4.5$

53. $-16.4-(-18.9+5.9)=-16.4-(-13)$
    $=-16.4+13=-3.4$

55. $510.65+279.19 \approx 510+280=790$

57. $671.01-88.35 \approx 700-90=610$

TRY IT YOURSELF

59. $-45.6+34.7=-10.9$

61. $-9.5-7.1=-9.5+(-7.1)=-16.6$

63. $46.09+(-7.8)=38.29$

65.   21.88
     +33.12

      55.00

67. $30.03-(-17.88)=30.03+17.88=47.91$

69. $645+9.90005+0.12+3.02002=658.04007$

71.   24.00
     −23.81

       0.19

73. $(3.4-6.6)+7.3=-3.2+7.3=4.1$

75. $247.9+40+0.56=288.46$

77.   78.10
      −7.81

      70.29

79. $-7.8+(-6.5)=-14.3$

81. $16-(67.2+6.27)=16-73.47=-57.47$

83. $2.43+5.6=8.03$

85. $|-14.1+6.9|+8=|-7.2|+8$
    $=7.2+8=15.2$

87. $5-0.023=4.977$

89. $-2.002-(-4.6)=-2.002+4.6=2.598$

APPLICATIONS

91. Refrigerator: $610.80 + $205.00 = $815.80

Washing Machine: $389.50 + 155.50 = $545.00

Dryer: $363.99 + $167.50 = $531.49

93. Design 1: 1.74 mi, 2.32 mi, 1.47 + 2.32 = 4.06 mi

Design 2: 2.90 mi, 0 mi, 2.90 + 0 = 2.90 mi

95. $0.218 + 0.218 + 1.939 = 2.375 in$

97. $52.88 - 10.49 = 42.39 \sec$

99. Subtotal: $523.19

Total Deposit: $523.19 - $25.00 = $498.19

101. Monday: 99.7° - 98.6° = 1.1° above normal

Tuesday: 98.6° + 2.5° = 101.1° in the A.M.

Wednesday: 98.6° - 98.6° = 0° above normal

Thursday: 100.0° - 98.6° = 1.4° above normal

Friday: 98.6° + 0.9° = 99.5° in the A.M.

103. $14.57 + 9.65 + 16.18 = 40.4 mi$

$40.4 - 20.39 = 20.01 mi$

105. a. $\$149.79 - \$47.85 = \$101.94$

b. $\$47.85 + \$7.95 = \$55.80$

WRITING

107. Because we can only add together items with the same place value.

109. The 37 should be written as 37.00 so the decimals line up.

111. To subtract the 6 we need to borrow from one place to the left. Since it is empty we must go one further left and borrow a 1 from the hundredths place. This gives us a 10 in the thousandths place from which we can now borrow.

REVIEW

113.

a. $\dfrac{4}{5} + \dfrac{5}{12} = \dfrac{4}{5} \cdot \dfrac{12}{12} + \dfrac{5}{12} \cdot \dfrac{5}{5}$

$= \dfrac{48}{60} + \dfrac{25}{60} = \dfrac{73}{60} = 1\dfrac{13}{60}$

b. $\dfrac{4}{5} - \dfrac{5}{12} = \dfrac{4}{5} \cdot \dfrac{12}{12} - \dfrac{5}{12} \cdot \dfrac{5}{5}$

$= \dfrac{48}{60} - \dfrac{25}{60} = \dfrac{23}{60}$

c. $\dfrac{4}{5} \cdot \dfrac{5}{12} = \dfrac{20}{60} = \dfrac{1}{3}$

d. $\dfrac{4}{5} \div \dfrac{5}{12} = \dfrac{4}{5} \cdot \dfrac{12}{5} = \dfrac{48}{25} = 1\dfrac{23}{25}$

Section 4.3: Multiplying Decimals

VOCABULARY

1.
```
      3.4   ← factor
    ×2.6   ← factor
      204  ← partial product
      680  ← partial procuct
     8.84  ← product
```

CONCEPTS

*Note in this section the numbers in the upper right hand corner$^{(these)}$ represent the number of significant digits right of the decimal places. Observe that the number in the product is equal to the sum of the numbers in the factors.

3.

a.
3.8^1
$\times 0.6^1$
2.28^2

b.
1.79^2
$\times 8.1^1$
179
14320
14.499^3

c. $\begin{array}{r} 2.0^1 \\ \times\ \ 7 \\ \hline 14.0^1 \end{array}$

d. $\begin{array}{r} 0.013^3 \\ \times 0.02^2 \\ \hline .00026^5 \end{array}$

5. a. Negative * Negative = Positive

    b. Negative * Positive = Negative

NOTATION

7. a. 10, 100, 1,000, 10,000, 100,000

    b. 0.1, 0.01, 0.001, 0.0001, 0.00001

GUIDED PRACTICE

9. $4.8 \cdot 6.2$

    $\begin{array}{r} 4.8^1 \\ \times 6.2^1 \\ \hline 96 \\ 2880 \\ \hline 29.76^2 \end{array}$

11. $5.6(8.9)$

    $\begin{array}{r} 5.6^1 \\ \times 8.9^1 \\ \hline 504 \\ 4480 \\ \hline 49.84^2 \end{array}$

13. $0.003(2.7)$

    $\begin{array}{r} 2.7^1 \\ \times 0.003^3 \\ \hline 0.0081^4 \end{array}$

15. $\begin{array}{r} 5.8^1 \\ \times 0.009^3 \\ \hline 0.0522^4 \end{array}$

17. $179(6.3)$

    $\begin{array}{r} 179^0 \\ \times 6.3^1 \\ \hline 537 \\ +10740 \\ \hline 1127.7^1 \end{array}$

19. $\begin{array}{r} 316^0 \\ \times 7.4^1 \\ \hline 1264 \\ +22120 \\ \hline 2338.4^1 \end{array}$

21. $6.84 \cdot 100$: 2 right : 684

23. $0.041(10,000)$: 4 right: 410

25. $647.59 \cdot 0.01$: 2 left : 6.4759

27. $1.15(0.001)$: 3 left : 0.00115

29. 14.2 million : 6 zeros = 6 right : 14,200,000

31. 98.2 billion : 9 zeros = 9 right : 98,200,000,000

33. 1.421 trillion : 12 zeros = 12 right : 1,421,000,000,000

35. 657.1 billion : 9 zeros = 9 right : 657,100,000,000

37. $-1.9(7.2)$

    $\begin{array}{r} 1.9^1 \\ \times 7.2^1 \\ \hline 38 \\ +1330 \\ \hline 13.68^2 \end{array}$

    $(-)(+) = -$, so

    -13.68

39.
$-3.3(-1.6)$

$$\begin{array}{r} 3.3^1 \\ \times 1.6^1 \\ \hline 198 \\ +330 \\ \hline 5.28^2 \end{array}$$

$(-)(-) = +,$ so
5.28

41. $(-10,000)(-44.83)$
$(-)(-) = +,$ so $10,000(44.83)$
4 right: $448,300$

43. $678.231(-1,000)$
$(+)(-) = -,$ so
3 right $= 678,231$
$-678,231$

45. $(3.4)^2 = (3.4)(3.4) = 11.56$

47. $(-0.03)^2 = (-0.03)(-0.03) = 0.0009$

49.
$-(-0.2)^2 + 4|-2.3+1.5|$
$= -(-0.2)^2 + 4|-0.8|$
$= -0.04 + 4(0.8)$
$= -0.04 + 3.2$
$= 3.16$

51.
$-(-0.8)^2 + 7|-5.1-4.8|$
$= -(-0.8)^2 + 7|-9.9|$
$= -0.64 + 7(9.9)$
$= -0.64 + 69.3$
$= 68.66$

53. $A = 85.50 + 85.50 \cdot 0.08 \cdot 5$
$= 85.50 + 34.2$
$= 119.70$

55. $A = (5.3)(7.2) = 38.16$

57. $P = 2(3.7) + 2(3.6) = 7.4 + 7.2 = 14.6$

59. $C = 2(3.14)(2.5) = 15.7$

61. $46 \cdot 5.3 \approx 50 \cdot 5 = 250$

63. $17.11 \cdot 3.85 \approx 17.1 \cdot 3.9 = 66.69$

TRY IT YOURSELF

65.
$-.56 \cdot 0.33$

$$\begin{array}{r} 56^2 \\ \times 33^2 \\ \hline 168 \\ 1680 \\ \hline .1848^4 \end{array}$$

$(-)(+) = -,$ so
-0.1848

67. $(-1.3)^2 = (-1.3)(-1.3) = 1.69$

69. $(-0.7-0.5)(2.4-3.1)$
$= (-1.2)(-0.7)$
$= 0.84$

71.
$$\begin{array}{cc} 0.008 & 8^3 \\ \times 0.09 & = \times 9^2 \\ \hline 0.00072^5 & \end{array}$$

73. $-0.2 \cdot 1,000,000$: 6 zeros, 6 right: $-200,000$

75. $(-5.6)(-2.2)$
$(5.6)(2.2) = 12.32$
$(-)(-) = (+),$ so
12.32

77. $-4.6(23.4-19.6)$
$= -4.6(3.8)$
$= -17.48$

79. $(-4.9)(-0.001)$: 3 left : 0.0049

81. $(-0.2)^2 + 2(7.1)$
$= 0.04 + 14.2$
$= 14.24$

83.
```
    2.13²
   ×4.05²
   ─────
    1065
       0
  +85200
  ──────
  8.6265⁴
```

85. $(-7)(8.1781)$
$(7)(8.1781) = 57.2467$
$(-)(+) = (-)$, so
-57.2467

87. $-1{,}000(0.02239)$: 3 zeros, 3 right: -22.39

89. $(0.5 + 0.6)^2(-3.2)$
$= (1.1)^2(-3.2)$
$= (1.21)(-3.2)$
$= -3.872$

91. $-0.2(306)(-0.4)$
$= (-61.2)(-0.4)$
$= 24.28$

93. $-0.01(|-2.6 - 6.7|)^2$
$= -0.01(|-9.3|)^2$
$= -0.01(9.3)^2$
$= -0.01(86.49)$
$= -0.8649$

95.

Decimal	Its square
0.1	0.01
0.2	0.04
0.3	0.09
0.4	0.16
0.5	0.25
0.6	0.36
0.7	0.49
0.8	0.64

APPLICATIONS

97. $500 \cdot (0.0038) = 1.9 in.$

99. $\$37.35 \cdot 38 \cdot 52.2 = 74087.46 \approx \$74{,}100$

101.

Type of nut	Price per Pound	Pounds	Cost
Almonds	$5.95	16	$95.20
Walnuts	$4.95	25	$123.75

103. $34 \overset{\circ}{A} = 34(0.000000004)$
$= 0.000000136 in.$

$3.4 \overset{\circ}{A} = 3.4(0.000000004)$
$= 0.0000000136 in.$

$10 \overset{\circ}{A} = 10(0.000000004)$
$= 0.00000004 in.$

105. a. $6 \cdot 0.35 = 2.1 mi.$

b. $10 \cdot 0.35 = 3.5 mi.$

c. $2.1 mi. + 3.5 mi. = 5.6 mi.$

107. $719 \cdot 0.14277 \approx \$102.65$

109. a. 19,6000,000 acres.

b. 6,500,000,000 people

c. 3,026,000,000,000 miles

111. a. $8 \times 24 = 192 \, ft.^2$

b. $25.5 \times 8.75 = 223.125 \, ft^2$

c. $223.125 - 192 = 31.125 \, ft^2$

113.

Ticket type	Price	Number sold	Receipts
Floor	$12.50	1,000	$12,500
Balcony	$15.75	100	$1,575

b. $14,075

115. $2(45.5 + 20.5 + 2.2) = 2(68.2) = 136.4 \, lbs.$

117. $0.57 + 3(.09) = 0.57 + 0.27 = 0.84 \, in.$

WRITING

119. Count the total number of places to the right of each decimal and have that many places right of the decimal in the answer.

121. To multiply a decimal by a power of 10 that is greater than 1, move the decimal to the right the number of zeros. To multiply by a power of 10 that is less than 1, move the decimal left the same number of places as there are in the power of 10.

123. Because multiplying will move the decimal anyways, so don't bother lining them up.

REVIEW

125. $220 = 2 \cdot 110 = 2 \cdot 10 \cdot 11$
$= 2 \cdot 2 \cdot 5 \cdot 11 = 2^2 \cdot 5 \cdot 11$

127. $162 = 2 \cdot 81 = 2 \cdot 9 \cdot 9$
$= 2 \cdot 3 \cdot 3 \cdot 3 \cdot 3 = 2 \cdot 3^4$

Section 4.4: Dividing Decimals

VOCABULARY

1. $\text{divisor} \rightarrow 5 \overline{)15.85} \leftarrow \text{dividend}$, quotient $= 3.17$

CONCEPTS

3. a. $4 \overline{)21.04} = 5.26$

b. $3 \overline{)0.024} = 0.008$

5. a. 1 right: $13 \overline{)106.6}$

b. 2 right: $371 \overline{)1669.5}$

7. $\dfrac{10}{10}$

9. thousandths

11. a. To find the quotient of a decimal and 10, 100, 1,000, and so on, move the decimal point to the left the same number of places as there are zeros in the power of 10.

b. To find the quotient of a decimal and 0.1, 0.01, 0.001, and so on, move the decimal point to the right the same number of decimal places as there are in the power of 10.

NOTATION

13. The red arrows indicate that the decimal points of both the divisor and dividend were moved 2 units to the right so the divisor is now a whole number.

GUIDED PRACTICE

15. $6 \overline{)12.6} = 2.1$
$-12 \downarrow$
06
-6
0

Check: $2.1 \times 6 = 12.6$ ☺

17.
$$\begin{array}{r}9.2\\3\overline{)27.6}\\\underline{-27}\downarrow\\06\\\underline{-6}\\0\end{array}$$

Check: $\begin{array}{r}9.2\\\times 3\\\hline 27.6\end{array}$ ☺

19.
$$\begin{array}{r}4.27\\23\overline{)98.21}\\\underline{-92}\downarrow\\62\\\underline{-46}\downarrow\\161\\\underline{-161}\\0\end{array}$$

Check: $4.27 \cdot 23 = 98.21$ ☺

21.
$$\begin{array}{r}8.65\\37\overline{)320.05}\\\underline{-296}\downarrow\\240\\\underline{-222}\downarrow\\185\\\underline{-185}\\0\end{array}$$

Check: $8.65 \cdot 37 = 320.05$ ☺

23.
$$\begin{array}{r}3.35\\4\overline{)13.40}\\\underline{-12}\downarrow\\14\\\underline{-12}\downarrow\\20\\\underline{-20}\\0\end{array}$$

Check: $3.35 \cdot 4 = 13.40$ ☺

25.
$$\begin{array}{r}4.56\\5\overline{)22.80}\\\underline{-20}\downarrow\\28\\\underline{-25}\downarrow\\30\\\underline{-30}\\0\end{array}$$

Check: $4.56 \cdot 5 = 22.8$ ☺

27. $\dfrac{0.1932}{0.42} = \dfrac{19.32}{42}$

$$\begin{array}{r}0.46\\42\overline{)19.32}\\\underline{-16\;8}\\2\;52\\\underline{-2\;52}\\0\end{array}$$

Check: $0.42 \cdot 0.46 = 0.1932$ ☺

29.

$$0.29\overline{)0.1131} = 29\overline{)11.31}$$

$$\begin{array}{r} 0.39 \\ 29\overline{)11.31} \\ \underline{-8\ 7} \\ 2\ 61 \\ \underline{-2\ 61} \\ 0 \end{array}$$

Check: $0.39 \cdot 0.29 = 0.1131$ ☺

31.

$$\dfrac{11.83}{0.6} = 6\overline{)118.300}$$

$$\begin{array}{r} 19.716.... \\ 6\overline{)118.300} \\ \underline{-6} \\ 58 \\ \underline{-54} \\ 43 \\ \underline{-42} \\ 10 \\ \underline{-6} \\ 40 \\ \underline{-36} \\ 4 \end{array}$$

≈ 19.72

Check: $19.72 \cdot 0.6 \approx 11.83$ ☺

33.

$$\dfrac{17.09}{0.7} = 7\overline{)170.900}$$

$$\begin{array}{r} 24.414...\\ 7\overline{)170.900} \\ \underline{-14} \\ 30 \\ \underline{-28} \\ 29 \\ \underline{-28} \\ 10 \\ \underline{-7} \\ 30 \\ \underline{-28} \\ 2 \end{array}$$

≈ 24.41

Check: $24.41 \cdot 0.7 \approx 17.09$ ☺

35. $289.842 \div 72.1 \approx 280 \div 70 = 4$

37. $383.76 \div 7.8 \approx 400 \div 8 = 50$

39. $3{,}883.284 \div 48.12 \approx 4{,}000 \div 50 = 80$

41. $6.1\overline{)15{,}819.74} \approx 15{,}000 \div 5 = 3{,}000$

43. $451.78 \div 100$: 2 zeros = 2 left : 4.5178

45. $\dfrac{30.09}{10{,}000}$: 4 zeros = 4 left : $.003009$

47. $1.25 \div 0.1$: 1 place = 1 right : 12.5

49. $\dfrac{545.2}{0.001}$: 3 places = 3 right : $545{,}200$

51. $-110.336 \div 12.8$:

$$128 \overline{\smash{\big)}\ 1103.36}$$
$$\underline{-1024}$$
$$793$$
$$\underline{-768}$$
$$256$$
$$\underline{-256}$$
$$0$$

quotient: 8.62

$(-) \div (+) = (-)$

-8.62

53. $-91.304 \div (-22.6)$

$$226 \overline{\smash{\big)}\ 913.04}$$
$$\underline{-904}$$
$$90$$
$$\underline{-0}$$
$$904$$
$$\underline{-904}$$
$$0$$

quotient: 4.04

$(-) \div (-) = (+)$

4.04

55. $\dfrac{-20.3257}{-0.001}$

3 places = 3 right

$(-) \div (-) = (+)$

20,325.7

57. $0.003 \div (-100)$

2 zeros = 2 left

$(+) \div (-) = (-)$

-0.00003

59. $\dfrac{2(0.614) + 2.3854}{0.2 - 0.9}$

$= \dfrac{1.228 + 2.3854}{-0.7}$

$= \dfrac{3.6134}{-0.7}$

$= -5.162$

61. $\dfrac{5.409 - 3(1.8)}{(0.3)^2}$

$= \dfrac{5.409 - 5.4}{0.09}$

$= \dfrac{0.009}{0.09}$

$= 0.1$

63. $t = \dfrac{d}{r} = \dfrac{211.75}{60.5} = 3.5$

65. $r = \dfrac{d}{t} = \dfrac{219.375}{3.75} = 58.5$

TRY IT YOURSELF

67. $4.5 \overline{\smash{\big)}\ 11.97}$

$$45 \overline{\smash{\big)}\ 119.70}$$
$$\underline{-90}$$
$$297$$
$$\underline{-270}$$
$$270$$
$$\underline{-270}$$
$$0$$

quotient: 2.66

69. $\dfrac{75.04}{10}$: 1 zero = 1 left : 7.504

71. $8 \overline{\smash{\big)}\ 0.0360}$

$$\underline{-32}$$
$$40$$
$$\underline{-40}$$
$$0$$

quotient: 0.0045

73.
$$\begin{array}{r} 0.321 \\ 9\overline{)2.889} \\ \underline{-27} \\ 18 \\ \underline{-18} \\ 09 \\ \underline{-9} \\ 0 \end{array}$$

75.
$$\frac{-3(0.2)-2(3.3)}{30(0.4)^2}$$
$$=\frac{-.6-6.6}{30(0.16)}$$
$$=\frac{-7.2}{4.8}$$
$$=-1.5$$

77. $1.2202 \div (-0.01)$

2 places = 2 right

$(+) \div (-) = (-)$

-122.02

79. $-5.714 \div 2.4$
$$\begin{array}{r} 2.38 \\ 24\overline{)571.40} \\ \underline{-48} \\ 91 \\ \underline{-72} \\ 194 \\ \underline{-192} \\ 2 \end{array}$$
$-2.38 \approx -2.4$

81. $-39 \div (-4)$
$$\begin{array}{r} 9.75 \\ 4\overline{)39.00} \\ \underline{-36} \\ 30 \\ \underline{-28} \\ 20 \\ \underline{-20} \\ 0 \end{array}$$

83. $7.8915 \div .00001$

5 places = 5 right

789,150

85. $\dfrac{0.0102}{0.017}$
$$\begin{array}{r} 0.6 \\ 17\overline{)10.2} \\ \underline{-102} \\ 0 \end{array}$$

87. $12.24 \div 0.9$
$$\begin{array}{r} 13.6 \\ 9\overline{)122.4} \\ \underline{-9} \\ 32 \\ \underline{-27} \\ 54 \\ \underline{-54} \\ 0 \end{array}$$

89. $1000\overline{)34.8}$

3 zeros = 3 left

0.0348

91. $$\frac{40.7(3-8.3)}{0.4-0.61}$$
$$=\frac{40.7(-5.3)}{-0.21}$$
$$=\frac{-215.71}{-0.21}$$
$$\approx 1{,}027.19$$

93. $\dfrac{0.25}{1.6} = \dfrac{2.5}{16} = 0.15625$

APPLICATIONS

95. $14 \div 0.05 = 280$ slices

97. $60 \div 0.00003 = 2{,}000{,}000$ computations

99. $8.5 \div 0.017 = 500$ squeezes

101. $27.5 \div 2.5 = 11$ hours: 6:00 PM

103. $1407.1 + 1388.2 + 1440.4 + 1482.5 + 1491.8 + 1510.4 = 8720.4$
$8720.4 \div 6 = 1{,}453.4$ million

105. $0.219 + 0.233 + 0.204 + 0.297 + 0.202 = 1.155$
$1.155 \div 5 = 0.231$ sec

WRITING

107. Begin by moving the decimals of both the divisor and dividend so the divisor is a whole number, adding zeros if necessary. Then divide as if it were an integer problem, placing the decimal in the quotient directly above the decimal in the dividend.

109. In this case equivalent means they have the same answer.

111. Round to the nearest 10th to get 2.4 and 0.8 which gives 3 as an approximation.

REVIEW

113. $10 = 2 \cdot 5$
$25 = 5 \cdot 5$
a. GCF = 5
b. LCM = $2 \cdot 5 \cdot 5 = 50$

Section 4.5: Fractions and Decimals

VOCABULARY

1. A fraction and a decimal are said to be <u>equivalent</u> if they name the same number.

3. 0.75, 0.625, and 3.5 are examples of <u>terminating</u> decimals.

CONCEPTS

5. $\dfrac{7}{8}$ means $7 \div 8$.

7. A decimal point and two additional <u>zeros</u> were written to the right of 3.

9. Sometimes, when finding the decimal equivalent of a fraction, the division process never gives a remainder of 0. We call the resulting decimal a <u>repeating</u> decimal.

11. a. 0.38
    b. 0.212

NOTATION

13. a. $0.7 = \dfrac{7}{10}$
    b. $0.77 = \dfrac{77}{100}$

GUIDED PRACTICE

15. $\begin{array}{r} 0.5 \\ 2\overline{)1.0} \\ \underline{-10} \\ 0 \end{array}$

17.
$$\begin{array}{r}0.875\\8\overline{)7.000}\\\underline{-64}\downarrow\\60\\\underline{-56}\downarrow\\40\\\underline{-40}\\0\end{array}$$

19.
$$\begin{array}{r}0.55\\20\overline{)11.00}\\\underline{-100}\downarrow\\100\\\underline{-100}\\0\end{array}$$

21.
$$\begin{array}{r}2.6\\5\overline{)13.0}\\\underline{-10}\downarrow\\30\\\underline{-30}\\0\end{array}$$

23.
$$\begin{array}{r}0.5625\\16\overline{)9.0000}\\\underline{-80}\downarrow\\100\\\underline{-96}\downarrow\\40\\\underline{-32}\downarrow\\80\\\underline{-80}\\0\end{array}$$

25.
$$\begin{array}{r}0.53125\\32\overline{)17.0000}\\\underline{-160}\downarrow\\100\\\underline{-96}\downarrow\\40\\\underline{-32}\downarrow\\80\\\underline{-64}\downarrow\\160\\\underline{-160}\\0\end{array}$$

Since the original problem was negative, -0.53125

27. $\dfrac{3}{5} = \dfrac{3}{5} \cdot \dfrac{2}{2} = \dfrac{6}{10} = 0.6$

29. $\dfrac{9}{40} = \dfrac{9}{40} \cdot \dfrac{25}{25} = \dfrac{225}{1000} = 0.225$

31. $\dfrac{19}{25} = \dfrac{19}{25} \cdot \dfrac{4}{4} = \dfrac{76}{100} = 0.76$

33. $\dfrac{1}{500} = \dfrac{1}{500} \cdot \dfrac{2}{2} = \dfrac{2}{1000} = 0.002$

35.
$$\begin{array}{r}0.75\\4\overline{)3.00}\\\underline{-28}\\20\\\underline{-20}\\0\end{array}$$

so $3\dfrac{3}{4} = 3.75$

37.

$$16\overline{)11.0000}$$
$$\underline{-96}$$
$$140$$
$$\underline{-128}$$
$$120$$
$$\underline{-112}$$
$$80$$
$$\underline{-80}$$
$$0$$

so $12\dfrac{11}{16} = 12.6875$

39.

$$9\overline{)1.00}^{\,0.11\cdots}$$
$$\underline{-9}$$
$$10$$
$$\underline{-9}$$
$$1$$

so $\dfrac{1}{9} = 0.\overline{1}$

41.

$$12\overline{)7.0000}^{\,0.5833\cdots}$$
$$\underline{-60}$$
$$100$$
$$\underline{-96}$$
$$40$$
$$\underline{-36}$$
$$40$$

$\dfrac{7}{12} = .58333\ldots = 0.58\overline{3}$

43.

$$90\overline{)7.000}^{\,0.077\cdots}$$
$$\underline{-630}$$
$$700$$
$$\underline{-630}$$
$$700$$

$\dfrac{7}{90} = .0777\ldots = 0.0\overline{7}$

45.

$$60\overline{)1.0000}^{\,0.0166\cdots}$$
$$\underline{-60}$$
$$400$$
$$\underline{-360}$$
$$400$$
$$\underline{-360}$$
$$400$$

$\dfrac{1}{60} = .016666\ldots = 0.01\overline{6}$

47.

$$11\overline{)5.000}^{\,0.454\cdots}$$
$$\underline{-44}$$
$$60$$
$$\underline{-55}$$
$$50$$
$$\underline{-44}$$
$$60$$

$-\dfrac{5}{11} = -.454545\ldots = -0.\overline{45}$

49.

$$33\overline{)20.000}^{\,0.606\cdots}$$
$$\underline{-198}$$
$$20$$
$$\underline{-0}$$
$$200$$
$$\underline{-198}$$
$$20$$

$-\dfrac{20}{33} = -.606060\ldots = -0.\overline{60}$

51.

$$\begin{array}{r} 0.233 \\ 30\overline{)7.000} \\ \underline{-60} \\ 100 \\ \underline{-90} \\ 100 \\ \underline{-90} \\ 10 \end{array}$$

$$\frac{7}{30} = 0.2\overline{3} \approx 0.23$$

53.

$$\begin{array}{r} 0.488 \\ 45\overline{)22.000} \\ \underline{-180} \\ 400 \\ \underline{-360} \\ 400 \\ \underline{-360} \\ 40 \end{array}$$

$$\frac{22}{45} = 0.4\overline{8} \approx 0.49$$

55.

$$\begin{array}{r} 1.846 \\ 13\overline{)24.000} \\ \underline{-13} \\ 110 \\ \underline{-104} \\ 60 \\ \underline{-52} \\ 80 \\ \underline{-78} \\ 2 \end{array}$$

$$\frac{24}{13} = 1.\overline{846153} \approx 1.85$$

57.

$$\begin{array}{r} 1.083 \\ 12\overline{)13.000} \\ \underline{-12} \\ 10 \\ \underline{-0} \\ 100 \\ \underline{-96} \\ 40 \\ \underline{-36} \\ 4 \end{array}$$

$$-\frac{13}{12} = -1.08\overline{3} \approx -1.08$$

59.

$$\begin{array}{r} 0.1515 \\ 33\overline{)5.0000} \\ \underline{-33} \\ 170 \\ \underline{-165} \\ 50 \\ \underline{-33} \\ 17 \end{array}$$

$$\frac{5}{33} = 0.151515\ldots \approx 0.152$$

61.

$$\begin{array}{r} 0.3703 \\ 27\overline{)10.0000} \\ \underline{-81} \\ 190 \\ \underline{-189} \\ 10 \\ \underline{-0} \\ 100 \\ \underline{-81} \\ 19 \end{array}$$

$$\frac{10}{27} = 0.370370370\ldots \approx 0.370$$

63.

65.

67. $\dfrac{7}{8} \square 0.895$

$0.875 \square 0.895$

$0.875 < 0.895$

69. $0.\overline{7} \square \dfrac{17}{22}$

$0.777... \square 0.7727...$

$0.777... > 0.7727...$

71. $\dfrac{52}{25} \square 2.08$

$2.08 \square 2.08$

$2.08 = 2.08$

73. $-\dfrac{11}{20} \square -0.\overline{48}$

$-.55 \square -0.4848...$

$-.55 < -0.4848...$

75. $6\dfrac{1}{2}, 6.25, \dfrac{19}{3}$

$= 6.5, 6.25, 6.\overline{3}$

so: $6.25, \dfrac{19}{3}, 6\dfrac{1}{2}$

77. $-0.\overline{81}, -\dfrac{8}{9}, -\dfrac{6}{7}$

$= -0.\overline{81}, -0.888..., -0.857...$

so: $-\dfrac{8}{9}, -\dfrac{6}{7}, -0.\overline{81}$

79. $\dfrac{1}{9} + 0.3 = \dfrac{1}{9} + \dfrac{3}{10} = \dfrac{10}{90} + \dfrac{27}{90} = \dfrac{37}{90}$

81. $0.9 - \dfrac{7}{12} = \dfrac{9}{10} - \dfrac{7}{12} = \dfrac{54}{60} - \dfrac{35}{60} = \dfrac{19}{60}$

83. $\dfrac{5}{11}(0.3) = \dfrac{5}{11} \cdot \dfrac{3}{10} = \dfrac{15}{110} = \dfrac{3}{22}$

85. $\dfrac{1}{4}(0.25) + \dfrac{15}{16} = \dfrac{1}{4} \cdot \dfrac{25}{100} + \dfrac{15}{16}$

$= \dfrac{25}{400} + \dfrac{375}{400} = \dfrac{400}{400} = 1$

87. $0.24 + \dfrac{1}{3} \approx 0.24 + 0.33 = 0.57$

89. $5.69 - \dfrac{5}{12} \approx 5.69 - 0.42 = 5.27$

91. $0.43 - \dfrac{1}{12} \approx 0.43 - 0.08 = 0.35$

93. $\dfrac{1}{15} - 0.55 \approx 0.07 - 0.55 = -0.48$

95. $(3.5 + 6.7)\left(-\dfrac{1}{4}\right)$

$= (10.2)(-0.25)$

$= -2.55$

97. $\left(\dfrac{1}{5}\right)^2 (1.7)$

$= 0.04(1.7)$

$= 0.068$

99. $7.5 - (0.78)\left(\dfrac{1}{2}\right)^2$

$= 7.5 - (0.78)(0.25)$

$= 7.5 - .195$

$= 7.305$

101.
$$\frac{3}{8}(3.2)+\left(4\frac{1}{2}\right)\left(-\frac{1}{4}\right)$$
$$=(0.375)(3.2)+(4.5)(-.25)$$
$$=1.2-1.125$$
$$=0.075$$

APPLICATIONS

103. Each line represents 0.0625 of an inch.

$0.0625, 0.375, 0.5625, 0.9375$

105. $3 \div 40 = 0.075$ - this is the thicker line.

107. $23.4 \sec, 23.8 \sec, 24.2 \sec, 32.6 \sec$

109. $6 \cdot \left(\frac{1}{2} \cdot 6 \cdot 5.2\right) = 6(15.6) = 93.6 in^2$

111. $\frac{2}{3} \cdot 4.14 + \frac{3}{4} \cdot 5.68 = \$7.02$

WRITING

113. To write a fraction as a decimal, use long division to divide the numerator by the denominator.

115. This is not the best form – the number of things under the bar should be minimized to $0.1\overline{3}$. Also, the 1 should not be under the bar.

117. By writing 0.76 as $\frac{76}{100}$ and then reducing.

REVIEW

119. a. $\{0,1,2,3,4,5,6,7,8,9\}$

b. $\{2,3,5,7,11,13,17,19,23,29\}$

c. $\{...,-3,-2,-1,0,1,2,3,...\}$

Section 4.6: Square Roots

VOCABULARY

1. When we raise a number to the second power, we are squaring it, or finding its square.

3. The symbol $\sqrt{\ }$ is called a radical symbol.

5. Whole numbers such as 36 and 49, that are squares of whole numbers, are called perfect squares.

CONCEPTS

7. a. The square of 5 is 25, because $5^2 = 25$.

b. The square of $\frac{1}{4}$ is $\frac{1}{16}$, because $\left(\frac{1}{4}\right)^2 = \frac{1}{16}$.

9. a. $\sqrt{49} = 7$ because $7^2 = 49$.

b. $\sqrt{4} = 2$ because $2^2 = 4$.

11. a. $\sqrt{1} = 1$

b. $\sqrt{0} = 0$

13. Step 2 – after dealing with grouping symbols.

15.

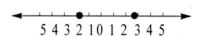

NOTATION

17. a. The symbol $\sqrt{\ }$ is used to indicate a positive square root.

b. The symbol $-\sqrt{\ }$ is used to indicate the negative square root of a positive number.

19. $-\sqrt{49} + \sqrt{64} = -7 + 8$
$= 1$

GUIDED PRACTICE

21. Since $5^2 = (-5)^2 = 25$ the square roots are $5, -5$.

23. Since $4^2 = (-4)^2 = 16$ the square roots are $4, -4$.

25. $\sqrt{16} = 4$ since $4^2 = 16$.

27. $\sqrt{9} = 3$ since $3^2 = 9$.

29. $-\sqrt{144} = -12$

31. $-\sqrt{49} = -7$

33. $\sqrt{961} = 31$

35. $\sqrt{3,969} = 63$

37. $\sqrt{\dfrac{4}{25}} = \dfrac{2}{5}$ since $\left(\dfrac{2}{5}\right)^2 = \dfrac{4}{25}$.

39. $-\sqrt{\dfrac{16}{9}} = -\dfrac{4}{3}$

41. $-\sqrt{\dfrac{1}{81}} = -\dfrac{1}{9}$

43. $\sqrt{0.64} = 0.8$ since $(0.8)^2 = 0.64$.

45. $-\sqrt{0.81} = -0.9$

47. $\sqrt{0.09} = 0.3$

49. $\sqrt{36} + \sqrt{1} = 6 + 1 = 7$

51. $\sqrt{81} + \sqrt{49} = 9 + 7 = 16$

53. $-\sqrt{144} - \sqrt{16} = -12 - 4 = -16$

55. $-\sqrt{225} + \sqrt{144} = -15 + 12 = -3$

57. $4\sqrt{25} = 4 \cdot 5 = 20$

59. $-10\sqrt{196} = -10 \cdot 14 = -140$

61. $-4\sqrt{169} + 2\sqrt{4} = -4 \cdot 13 + 2 \cdot 2$
$= -52 + 4 = -48$

63. $-8\sqrt{16} + 5\sqrt{225} = -8 \cdot 4 + 5 \cdot 15$
$= -32 + 75 = 43$

65. $15 + 4\left[5^2 - (6-1)\sqrt{4}\right]$
$= 15 + 4\left[25 - (5)2\right]$
$= 15 + 4\left[25 - 10\right]$
$= 15 + 4\left[15\right]$
$= 15 + 60$
$= 75$

67. $50 - \left[(6^2 - 24) + 9\sqrt{25}\right]$
$= 50 - \left[(36 - 24) + 9 \cdot 5\right]$
$= 50 - \left[12 + 45\right]$
$= 50 - 57$
$= -7$

69. $\sqrt{196} + 3\left(5^2 - 2\sqrt{225}\right)$
$= 14 + 3(25 - 2 \cdot 15)$
$= 14 + 3(25 - 30)$
$= 14 + 3(-5)$
$= 14 + (-15)$
$= -1$

71. $\dfrac{\sqrt{16} - 6(2^2)}{\sqrt{4}}$
$= \dfrac{4 - 6(4)}{2}$
$= \dfrac{4 - 24}{2}$
$= \dfrac{-20}{2}$
$= -10$

73. $\sqrt{\dfrac{1}{16}} - \sqrt{\dfrac{9}{25}} = \dfrac{1}{4} - \dfrac{3}{5} = \dfrac{5}{20} - \dfrac{12}{20} = -\dfrac{7}{20}$

75. $5\left(-\sqrt{49}\right)(-2)^2 = 5(-7)(4) = -140$

77. $(6^2)\sqrt{0.04} + 2.36 = 36 \cdot 0.2 + 2.36$
$= 7.2 + 2.36 = 9.56$

79. $-\left(-3\sqrt{1.44}+5\right)$
$=-\left(-3 \cdot 1.2+5\right)$
$=-\left(-3.6+5\right)$
$=-1.4$

81. $c=\sqrt{9^2+12^2}=\sqrt{81+144}=\sqrt{225}=15$

83. $a=\sqrt{25^2-24^2}=\sqrt{625-576}=\sqrt{49}=7$

85.
Number	Square Root
1	1
2	1.414
3	1.732
4	2
5	2.236
6	2.449
7	2.646
8	2.828
9	3
10	3.162

87. $\sqrt{15}\approx 3.87$

89. $\sqrt{66}\approx 8.12$ (Get your kicks!)

91. $\sqrt{24.05}\approx 4.904$

93. $-\sqrt{11.1}\approx -3.332$

APPLICATIONS

95. a. $\sqrt{25}\,ft.=5\,ft.$

    b. $\sqrt{100}\,ft.=10\,ft.$

97. $\sqrt{16,200}\approx 127.3\,ft.$

99. $\sqrt{1,764}\,in.=42\,in.$ - a 42" screen.

WRITING

101. He is dividing the number by 2, rather than finding the number that will give 16 when raised to the second power.

103. A non-terminating decimal is one where division will never produce a remainder of 0. The decimal obtained from dividing 1 by 6 is an example of this.

105. It does not represent a real number because the square of any real number will always be a real number so it is not possible to have a real square root of a negative number.

107. Because 2.449 is an approximation since the actual square root is a non-terminating decimal.

REVIEW

109. $6.75 \cdot 12.2 = 82.35$

111. $(3.4)^3 = (3.4)(3.4)(3.4) = 39.304$

Chapter 4 Review

1. a. $0.67;\dfrac{67}{100}$

   b.  0.8

3. $10+6+\dfrac{4}{10}+\dfrac{5}{100}+\dfrac{2}{1,000}+\dfrac{3}{10,000}$

5. negative six hundred fifteen and fifty-nine hundredths

   $-615\dfrac{59}{100}$

7. one hundred-thousandth; $\dfrac{1}{100,000}$

9. 11.997

11. $5.68 < 5.75$

13. $-78.23 > -78.303$

15.

17. 4.58

19. −0.1

21. 6.7030

23. 0.222228

25. $0.67

27. Washington, Diaz, Chou, Singh, Gerbac

29. 19.5
    34.4
    +12.8
    66.7

31. 68.47
    −53.30
    15.17

33. 9,000.090
    −7,067.445
    1,932.645

35. Subtract the smaller absolute value from the larger. Since the negative number has the larger absolute value, the answer will be negative.
    $16.1 - 8.4 = 7.7$ so $-16.1 + 8.4 = -7.7$

37. Add the absolute values. Since both numbers are negative, the result will be negative.
    $3.55 + 1.25 = 4.8$ so
    $-3.55 + (-1.25) = -4.8$

39. $-8.8 + (-7.3 - 9.5)$
    $= -8.8 + (-16.8)$
    $= -25.6$

41. a. $610 + 150 = 760$
    b. $300 - 20 = 280$

43. 52.20
    − 3.99
    $48.21

45. 2.3
    ×6.9
    207
    +1380
    15.87

47. 1.7
    ×0.004
    0.0068

49. 15.5
    ×9.8
    1240
    +13950
    151.90

    so $15.5(-9.8) = -151.9$

51. Since there are 3 zeros after the 1 in 1,000, move the decimal point 3 units to the right.
    $1,000(90.1452) = 90,145.2$

53. $(0.2)^2 = (0.2)(0.2) = 0.04$

55. $(0.6 + 0.7)^2 - (-3)(-4.1)$
    $= (1.3)^2 - (12.3)$
    $= 1.69 - 12.3$
    $= -10.61$

57. $(-3.3)^2(0.00001)$
$= 10.89(0.00001)$
$= 0.0001089$

59. a. $9,600,000 \, km^2$

b. $2,310,000,000$ trees

61. $A = 70.05 + 70.05 \cdot 0.08 \cdot 5$
$= 70.05 + 5.604 \cdot 5$
$= 70.05 + 28.02$
$= 98.07$

63. $0.03 + 3(0.015) - 0.005$
$= 0.03 + 0.045 - 0.005$
$= 0.075 - 0.005$
$= 0.07 \, in$

65.
$$\begin{array}{r} 9.3 \\ 3\overline{)27.9} \\ \underline{-27} \\ 09 \\ \underline{-9} \\ 0 \end{array}$$

Check: $3(9.3) = 27.9$

67.
$$\begin{array}{r} 1.29 \\ 23\overline{)29.67} \\ \underline{-23} \\ 66 \\ \underline{-46} \\ 207 \\ \underline{-207} \\ 0 \end{array}$$

Check: $23(1.29) = 29.67$

69. $-80.625 \div 12.9 = -806.25 \div 129$

$$\begin{array}{r} 6.25 \\ 129\overline{)806.25} \\ \underline{-774} \\ 322 \\ \underline{-258} \\ 645 \\ \underline{-645} \\ 0 \end{array}$$

Since (neg)/(pos)=(neg), our answer is -6.25

Check: $-6.25(12.9) = -80.625$

71. $\dfrac{15.75}{0.25} = \dfrac{1575}{25}$

$$\begin{array}{r} 63 \\ 25\overline{)1575} \\ \underline{-150} \\ 75 \\ \underline{-75} \\ 0 \end{array}$$

Check: $0.25(63) = 15.75$

73. Since there are 3 zeros right of the 1 in 1,000, move the decimal 3 units left.

$89.76 \div 1,000 = 0.08976$

Check: $1,000(0.08976) = 89.76$

75. Since 0.001 has 3 decimal places, move the decimal 3 units right and make the result positive since we are dividing a negative by a negative.

$-0.8765 \div (-0.001) = 876.5$

Check: $876.5(-0.001) = -0.8765$

77. $4,800 \div 40 = 480 \div 4 = 120$

79. Start by moving each decimal 1 to the right.

$$
\begin{array}{r}
12.94 \\
61\overline{)789.80} \\
\underline{-61} \\
179 \\
\underline{-122} \\
578 \\
\underline{-549} \\
290
\end{array}
$$

To the nearest 10th, 12.9

81. $\dfrac{(1.4)^2 - 2(-4.6)}{0.5 + 0.3}$

$= \dfrac{1.96 + 9.2}{0.8}$

$= \dfrac{11.16}{0.8}$

$= \dfrac{111.6}{8}$

$= 13.95$

83.
$$
\begin{array}{r}
8.34 \\
5\overline{)41.70} \\
\underline{-40} \\
17 \\
\underline{-15} \\
20 \\
\underline{-20} \\
0
\end{array}
$$

$8.34 per person

85. $\dfrac{15.4}{1.1} = \dfrac{154}{11}$

$$
\begin{array}{r}
14 \\
11\overline{)154} \\
\underline{-14} \\
14 \\
\underline{-14} \\
0
\end{array}
$$

14 servings per container

87.
$$
\begin{array}{r}
0.875 \\
8\overline{)7.000} \\
\underline{-64} \\
60 \\
\underline{-56} \\
40 \\
\underline{-40} \\
0
\end{array}
$$

$\dfrac{7}{8} = 0.875$

89.
$$
\begin{array}{r}
0.5625 \\
16\overline{)9.0000} \\
\underline{-80} \\
100 \\
\underline{-96} \\
40 \\
\underline{-32} \\
80 \\
\underline{-80} \\
0
\end{array}
$$

$\dfrac{9}{16} = 0.5625$

91.
$$
\begin{array}{r}
.5454 \\
11\overline{)6.0000} \\
\underline{-55} \\
50 \\
\underline{-44} \\
60 \\
\underline{-55} \\
50 \\
\underline{-44} \\
6
\end{array}
$$

$\dfrac{6}{11} = 0.\overline{54}$

93.
$$125\overline{)7.000}$$
$\underline{-625}$
750
$\underline{-750}$
0

$3\dfrac{7}{125} = 3.056$

95.
$$33\overline{)19.000}$$
$\underline{-165}$
250
$\underline{-231}$
190

$\dfrac{19}{33} \approx 0.58$

97. $\dfrac{13}{25} \,\square\, 0.499$

$0.52 > 0.499$

99. $\dfrac{10}{33} = .\overline{30}$

$0.3, \dfrac{10}{33}, 0.\overline{3}$

101. $\dfrac{1}{3} + 0.4 = \dfrac{1}{3} + \dfrac{4}{10} = \dfrac{1}{3} + \dfrac{2}{5}$

$= \dfrac{5}{15} + \dfrac{6}{15} = \dfrac{11}{15}$

103. $\dfrac{4}{5}(-7.8) = 0.8(-7.8) = -6.24$

105. $\dfrac{1}{2}(9.7 + 8.9)(10) = 0.5(18.6)(10)$

$= 9.3(10) = 93$

107. $A = \dfrac{1}{2}bh = 0.5(6.4)(10.9)$

$= 3.2(10.9) = 34.88 \, in^2$

109. $-5, 5$

111. $\sqrt{49} = 7$

113. $\sqrt{100} = 10$

115. $\sqrt{\dfrac{64}{169}} = \dfrac{8}{13}$

117. $-\sqrt{\dfrac{1}{36}} = -\dfrac{1}{6}$

119.

121. $-3\sqrt{100} = -3 \cdot 10 = -30$

123. $-3\sqrt{49} - 6 = -3 \cdot 7 - 6$

$= -21 - 6 = -27$

125. $40 + 6\left[5^2 - (7-2)\sqrt{16}\right]$

$= 40 + 6[25 - 5 \cdot 4]$

$= 40 + 6[25 - 20]$

$= 40 + 6[5]$

$= 40 + 30$

$= 70$

127. $b = \sqrt{c^2 - a^2} = \sqrt{17^2 - 15^2}$

$= \sqrt{289 - 225} = \sqrt{64} = 8$

129. $\sqrt{81} < \sqrt{83} < \sqrt{100}$

$9 < \sqrt{83} < 10$

Chapter 4 Test

1. a. addend

    addend

    sum

    b. minuend

    subtrahend

    difference

    c. factor

    factor

    product

    d.   quotient

    divisor   dividend

    e. repeating

    f. radical

3. a. 1 thousandth

    b. 4

    c. 6

    d. 2 tens

5. 4,519.0027

7. a. 461.7

    b. 2,733.050

    c. -1.983383

9. $\overset{1\ 1\ 1}{4.560}$
    2.000
    0.896
    + 3.300
    ———
    10.756

11. $(0.32)^2 = (0.32)(0.32)$

    0.32^2
    $\times 0.32^2$
    ————
    64
    960
    ————
    0.1024^4

13. $-6.7(-2.1) = 6.7(2.1)$

    6.7^1
    $\times 2.1^1$
    ————
    67
    +1340
    ————
    14.07^2

15. $\quad\quad 1.181$
    $11\overline{)13.000}$
    $\quad\underline{-11}$
    $\quad\quad 20$
    $\quad\underline{-11}$
    $\quad\quad\quad 90$
    $\quad\quad\underline{-88}$
    $\quad\quad\quad\quad 20$

    $\dfrac{13}{11} = 1.\overline{18}$

17. $12.146 \div 5.3 = 121.46 \div 53$

    $\quad\quad 2.290$
    $53\overline{)121.460}$
    $\quad\underline{-106}$
    $\quad\quad 154$
    $\quad\quad\underline{-106}$
    $\quad\quad\quad 480$
    $\quad\quad\quad\underline{-477}$
    $\quad\quad\quad\quad 30$

    $\dfrac{121.46}{-5.3} \approx -2.29$

19. a. Move the decimal 3 left: 0.567909

    b. Move the decimal 2 right: 0.458

21.
$\overset{11}{0.834}$
$+0.192$
1.026

1.026 in

23. $2.26 \div 565$
$= 0.004$ in

25. $4.37 - 1.86 - 2.09 = C$
$2.51 - 2.09 = C$
$0.42 g = C$

27. $C = 2\pi r$
$C = 2 \cdot 3.14(1.7)$
$C = 6.28(1.7)$
$C = 10.676$

29. $4.1 - (3.2)(0.4)^2$
$= 4.1 - (3.2)(0.16)$
$= 4.1 - 0.512$
$= 3.588$

31. $8 - 2(2^4 - 60 + 6\sqrt{81})$
$= 8 - 2(16 - 60 + 6 \cdot 9)$
$= 8 - 2(-44 + 54)$
$= 8 - 2(10)$
$= 8 - 20$
$= -12$

33.
a.
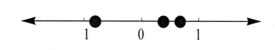

b.

35. $\sqrt{a^2 + b^2}$
$= \sqrt{12^2 + 35^2}$
$= \sqrt{144 + 1225}$
$= \sqrt{1369}$
$= 37$

37. $\sqrt{144} = 12$ because $12^2 = 144$.

39. $-2\sqrt{25} + 3\sqrt{49}$
$= -2 \cdot 5 + 3 \cdot 7$
$= -10 + 21$
$= 11$

41. a. $-\sqrt{0.04} = -0.2$
b. $\sqrt{1.69} = 1.3$
c. $\sqrt{225} = 15$
d. $-\sqrt{121} = -11$

Chapters 1 – 4 Cumulative Review

1. a. one hundred fifty-four thousand, three hundred two
   b. $100{,}000 + 50{,}000 + 4{,}000 + 300 + 2$

3.
$\overset{1111}{9{,}339}$
471
$+\ 6{,}883$
$16{,}693$

5.
$$\begin{array}{r}275\\ \times 275\\ \hline 1,375\\ 19,250\\ +55,000\\ \hline 75,625\,ft^2\end{array}$$

7.
$$\begin{array}{r}{\cancel{3}}{\cancel{9}}{\cancel{9}}{\cancel{10}}\\ \cancel{4}\,\cancel{0}\,\cancel{0}\,,\cancel{0}\,00\\ -2\;2\;1\;,1\;00\\ \hline 1\;7\;8\;,9\;00\end{array}$$

9. $220 = 2 \cdot 110 = 2 \cdot 2 \cdot 55$
  $= 2 \cdot 2 \cdot 5 \cdot 11 = 2^2 \cdot 5 \cdot 11$

11. $\dfrac{7+1+8+2+2}{5}$
   $= \dfrac{20}{5} = 4$

13. $-8 + (-5) = -13$

15. $72 - (-11) = 72 + 11 = 83°F$ increase

17. $(-1)^5 = (-1)(-1)(-1)(-1)(-1)$
   $= 1 \cdot 1(-1) = -1$

19. $\dfrac{-15}{-5} = 3 \Leftrightarrow 3(-5) = -15$

21. $(-6)^2 - 2(5 - 4 \cdot 2) = (-6)^2 - 2(5 - 8)$
   $= (-6)^2 - 2(-3)$
   $= 36 - 2(-3)$
   $= 36 - (-6)$
   $= 36 + 6$
   $= 42$

23. $-3,900 + (-5,800) + 4,700$
   $= -9,700 + 4,700$
   $= -5,000$

Chapter 4 Decimals 109

25. This illustrates the concept of equivalent fractions.

27. $\dfrac{3}{8} \cdot \dfrac{7}{16} = \dfrac{21}{128}$

29. $\dfrac{1}{9} + \dfrac{5}{6} = \dfrac{1}{9} \cdot \dfrac{2}{2} + \dfrac{5}{6} \cdot \dfrac{3}{3} = \dfrac{2}{18} + \dfrac{15}{18} = \dfrac{17}{18}$

31.
$$\begin{array}{r}76\dfrac{1}{6} = 76\dfrac{4}{24} = 75\dfrac{28}{24}\\ -49\dfrac{7}{8} = -49\dfrac{21}{24} = -49\dfrac{21}{24}\\ \hline 26\dfrac{7}{24}\end{array}$$

33. $\dfrac{1}{4} \cdot \dfrac{7}{16} = \dfrac{7}{64}$

35. $\left(\dfrac{9}{20} \div 2\dfrac{2}{5}\right) + \left(\dfrac{3}{4}\right)^2$
   $= \left(\dfrac{9}{20} \div \dfrac{12}{5}\right) + \left(\dfrac{3}{4}\right)\left(\dfrac{3}{4}\right)$
   $= \left(\dfrac{9}{20} \cdot \dfrac{5}{12}\right) + \left(\dfrac{9}{16}\right)$
   $= \left(\dfrac{\cancel{9}^3}{\cancel{20}_4} \cdot \dfrac{\cancel{5}}{\cancel{12}_4}\right) + \left(\dfrac{9}{16}\right)$
   $= \dfrac{3}{16} + \dfrac{9}{16} = \dfrac{12}{16} = \dfrac{3}{4}$

37. $356.1978 < 356.22$

39.
$$\begin{array}{r}{}^{1\;3\;1}\\ 56.228\\ 5.600\\ 39.000\\ +\;29.370\\ \hline 130.198\end{array}$$

41. $-1.8(4.52)$

$$\begin{array}{r} 4.52^2 \\ \times 1.8^1 \\ \hline 3616 \\ +4520 \\ \hline 8.136^3 \end{array}$$

$-1.8(4.52) = -8.146$

43. $\dfrac{-21.28}{-3.8} = \dfrac{212.8}{38}$

$$\begin{array}{r} 5.6 \\ 38\overline{)212.8} \\ -190 \\ \hline 228 \\ -228 \\ \hline 0 \end{array}$$

$\dfrac{-21.28}{-3.8} = 5.6$

45. $168 - 48.3 - 34.5 - 21.0 - 3.1 - 27.5$
 $= 119.7 - 34.5 - 21.0 - 3.1 - 27.5$
 $= 85.2 - 21.0 - 3.1 - 27.5$
 $= 64.2 - 3.1 - 27.5$
 $= 61.1 - 27.5$
 $= 33.6 hr$

47. $C = \dfrac{5}{9}(451 - 32)$

$C = \dfrac{5}{9}(419)$

$C = 232.\overline{7} \approx 232.8$

49. $\dfrac{3}{8}(-3.2) + \left(4\dfrac{1}{2}\right)\left(-\dfrac{1}{4}\right)$
 $= 0.375(-3.2) + (4.5)(-0.25)$
 $= -1.2 - 1.125 = -2.325$

51. $\sqrt{49} = 7$

53. $-4\sqrt{36} + 2\sqrt{81} = -4 \cdot 6 + 2 \cdot 9$
 $= -24 + 18 = -6$

CHAPTER 5: Ratio, Proportion, and Measurement

Section 5.1: Ratios

VOCABULARY

1. A <u>ratio</u> is the quotient of two numbers or the quotient of two quantities that have the same units.

3. A <u>unit</u> rate is a rate in which the denominator is 1.

CONCEPTS

5. 3 is a factor of both 15 and 24.

7. Multiplying numerator and denominator by 10 will give $\frac{5}{6}$.

9. $\frac{11 \text{ min}}{60 \text{ min}} = \frac{11}{60}$

NOTATION

11. $\frac{13}{9}$, 13 to 9, 13:9

GUIDED PRACTICE

13. $\frac{5}{8}$

15. $\frac{11}{16}$

17. $\frac{25}{15} = \frac{5 \cdot \cancel{5}}{3 \cdot \cancel{5}} = \frac{5}{3}$

19. $\frac{63}{36} = \frac{7 \cdot \cancel{9}}{4 \cdot \cancel{9}} = \frac{7}{4}$

21. $\frac{22}{33} = \frac{2 \cdot \cancel{11}}{3 \cdot \cancel{11}} = \frac{2}{3}$

23. $\frac{17}{34} = \frac{\cancel{17}}{2 \cdot \cancel{17}} = \frac{1}{2}$

25. $\frac{4 \text{ oz}}{12 \text{ oz}} = \frac{\cancel{4} \cancel{\text{oz}}}{3 \cdot \cancel{4} \cancel{\text{oz}}} = \frac{1}{3}$

27. $\frac{24 \text{ mi}}{32 \text{ mi}} = \frac{3 \cdot \cancel{8} \cancel{\text{mi}}}{4 \cdot \cancel{8} \cancel{\text{mi}}} = \frac{3}{4}$

29. $\frac{0.3}{0.9} = \frac{0.3}{0.9} \cdot \frac{10}{10} = \frac{3}{9} = \frac{1}{3}$

31. $\frac{0.65}{0.15} = \frac{0.65}{0.15} \cdot \frac{100}{100} = \frac{65}{15} = \frac{13}{3}$

33. $\frac{3.8}{7.8} = \frac{3.8}{7.8} \cdot \frac{10}{10} = \frac{38}{78} = \frac{19}{39}$

35. $\frac{7}{24.5} = \frac{7}{24.5} \cdot \frac{10}{10} = \frac{70}{245} = \frac{2}{7}$

37. $\frac{2\frac{1}{3}}{4\frac{2}{3}} = \frac{\frac{7}{3}}{\frac{14}{3}} = \frac{7}{3} \cdot \frac{3}{14} = \frac{\cancel{7}}{\cancel{3}} \cdot \frac{\cancel{3}}{2 \cdot \cancel{7}} = \frac{1}{2}$

39. $\frac{10\frac{1}{2}}{1\frac{3}{4}} = \frac{\frac{21}{2}}{\frac{7}{4}} = \frac{21}{2} \cdot \frac{4}{7} = \frac{3 \cdot \cancel{7}}{\cancel{2}} \cdot \frac{\cancel{2} \cdot 2}{\cancel{7}} = \frac{6}{1}$

41. $\frac{12 \text{ min}}{1 \text{ hr}} = \frac{12 \text{ min}}{60 \text{ min}} = \frac{\cancel{12} \cancel{\text{min}}}{5 \cdot \cancel{12} \cancel{\text{min}}} = \frac{1}{5}$

43. $\frac{3d}{1 \text{ wk}} = \frac{3d}{7d} = \frac{3\cancel{d}}{7\cancel{d}} = \frac{3}{7}$

45. $\frac{18 \text{ mo}}{2 \text{ yr}} = \frac{18 \text{ mo}}{24 \text{ mo}} = \frac{3 \cdot \cancel{6} \cancel{\text{mo}}}{4 \cdot \cancel{6} \cancel{\text{mo}}} = \frac{3}{4}$

47. $\frac{21 \text{ in}}{3 \text{ ft}} = \frac{21 \text{ in}}{36 \text{ in}} = \frac{\cancel{3} \cdot 7 \cancel{\text{in}}}{\cancel{3} \cdot 12 \cancel{\text{in}}} = \frac{7}{12}$

49. $\frac{64 \text{ ft}}{6 \text{ sec}} = \frac{\cancel{2} \cdot 32 \text{ ft}}{\cancel{2} \cdot 3 \text{ sec}} = \frac{32 \text{ ft}}{3 \text{ sec}}$

51. $\frac{75d}{20 \text{ gal}} = \frac{\cancel{5} \cdot 15d}{4 \cdot \cancel{5} \text{ gal}} = \frac{15d}{4 \text{ gal}}$

53. $\dfrac{84 \text{ made}}{100 \text{ attempts}} = \dfrac{\cancel{4} \cdot 21 \text{ made}}{\cancel{4} \cdot 25 \text{ attempts}} = \dfrac{21 \text{ made}}{25 \text{ attempts}}$

55. $\dfrac{18 \text{ beats}}{12 \text{ measures}} = \dfrac{3 \cdot \cancel{6} \text{ beats}}{2 \cdot \cancel{6} \text{ measures}} = \dfrac{3 \text{ beats}}{2 \text{ measures}}$

57. $\dfrac{60 \text{ rev}}{5 \text{ min}} = \dfrac{\cancel{5} \cdot 12 \text{ rev}}{\cancel{5} \text{ min}} = 12$ revolutions per minute

59. $\dfrac{\$50,000}{10 \text{ yr}} = \dfrac{\cancel{10} \cdot \$5,000}{\cancel{10} \text{ yr}} = \$5,000$ per year

61. $12 \div 8 = 1.5$

    1.5 errors per hour

63. $4,007,500 \div 12,500 = 320.6$

    320.6 people per square mile

65. $48 \div 12 = 4$

    $4 per minute

67. $272 \div 4 = 68$

    $68 each

69. $78 \div 65 = 1.2$

    1.2 cents per ounce

71. $3.50 \div 50 = 0.07$

    $0.07 per foot

APPLICATIONS

73. a. $\dfrac{12}{18} = \dfrac{2}{3}$

    b. $\dfrac{18}{12} = \dfrac{3}{2}$

75. $\dfrac{0.02}{1.1} = \dfrac{2}{110} = \dfrac{1}{55}$

77. $\dfrac{5\tfrac{1}{4}}{1\tfrac{3}{4}} = \dfrac{\tfrac{21}{4}}{\tfrac{7}{4}} = \dfrac{21}{4} \cdot \dfrac{4}{7} = \dfrac{3}{1}$

79. a. $800 + 600 + 100 + 100 + 120 + 80 = \$1,800$

    b. $\dfrac{800}{1800} = \dfrac{4}{9}$

    c. $\dfrac{600}{1800} = \dfrac{1}{3}$

    d. $\dfrac{100}{1800} = \dfrac{1}{18}$

81. Since the height and width of a square are the same, the ratio is $\dfrac{1}{1}$.

83. $\dfrac{5¢}{\$1} = \dfrac{5¢}{100¢} = \dfrac{1}{20}$

85. $\dfrac{125}{50} = \dfrac{5 \cdot 25}{2 \cdot 25} = \dfrac{5 \text{ compressions}}{2 \text{ breaths}}$

87. $\dfrac{3.29}{1,000} = \dfrac{329 \text{ complaints}}{100,000 \text{ passengers}}$

89. a. $112,500 - 4,500 = 108,000$

    b. $\dfrac{108,000}{4,500} = 24$ browsers per buyer

91. $\dfrac{84}{12} = 7¢/oz$

93. $\dfrac{25}{20} = 1.25¢/min$

95. $\dfrac{222.50}{50} = \$4.45/lb$

97. $\dfrac{11,880}{27} = 440$ gallons per minute

99. a. $35,071 - 34,746 = 325$ miles

b. $\dfrac{325}{5} = 65$ mph

101. The 6 oz. can: $\dfrac{89}{6} \approx 14.8¢/oz$

The 8 oz. can: $\dfrac{119}{8} \approx 14.9¢/oz$

So the 6 oz. can is a better buy.

103. The box of 20: $\dfrac{429}{20} = 21.45¢/\text{pill}$

The box of 50: $\dfrac{959}{50} = 19.18¢/\text{pill}$

So the 50-tablet box is a better buy.

105. Car: $\dfrac{345}{6} = 57.5$ mph

Truck: $\dfrac{376}{6.2} \approx 60.6$ mph

So the truck is traveling faster.

107. First car: $\dfrac{1,235}{51.3} \approx 24.1\, mpg$

Second car: $\dfrac{1,456}{55.78} \approx 26.1\, mpg$

So the second car had better mileage.

WRITING

109. No. In the first case the first quantity is triple the second while in the second case the first quantity is a third of the second.

111. It will be easier to figure out which sales are actually bargains.

REVIEW

113. $12,897 + 29,431 + 2,595$
$\approx 10,000 + 30,000 + 3,000$
$= 43,000$

115. $410 \cdot 21$
$\approx 400 \cdot 20$
$= 8,000$

Section 5.2: Proportions

VOCABULARY

1. A <u>proportion</u> is a statement that two ratios (or rates) are equal.

3. The <u>cross</u> products for the proportion $\dfrac{4}{7} = \dfrac{36}{x}$ are $4 \cdot x$ and $7 \cdot 36$.

5. A letter that is used to represent an unknown number is called a <u>variable</u>.

7. We solve proportions by writing a series of steps that result in an equation of the form $x = $ a number or a number $= x$. We say that the variable x is <u>isolated</u> on one side of the equation.

CONCEPTS

9. If the cross products of a proportion are equal, the proportion is <u>true</u>. If the cross products are *not* equal, the proportion is <u>false</u>.

11. $9 \cdot 10 = 90 \qquad 2 \cdot 45 = 90$

$\dfrac{9}{2} \bowtie \dfrac{45}{10}$

13. $\dfrac{\text{Teacher's aides} \to \ 12}{\text{Children} \to 100} = \dfrac{3}{25} \begin{matrix}\leftarrow \text{Teacher's aid} \\ \leftarrow \text{Children}\end{matrix}$

NOTATION

15. $2 \cdot 9 = 3 \cdot x$
$18 = 3 \cdot x$
$\dfrac{18}{3} = \dfrac{3 \cdot x}{3}$
$6 = x$

The solution is 6.

GUIDED PRACTICE

17. $\dfrac{20}{30} = \dfrac{2}{3}$

19. $\dfrac{400 \text{ sheets}}{100 \text{ beds}} = \dfrac{4 \text{ sheets}}{1 \text{ bed}}$

21. $\dfrac{70}{81} = \dfrac{2 \cdot 5 \cdot 7}{3 \cdot 3 \cdot 3 \cdot 3}$

    Nothing simplifies so the proportion is false.

23. $\dfrac{21}{14} = \dfrac{3 \cdot \cancel{7}}{2 \cdot \cancel{7}} = \dfrac{3}{2}$

    $\dfrac{18}{12} = \dfrac{3 \cdot \cancel{6}}{2 \cdot \cancel{6}} = \dfrac{3}{2}$

    The proportion is true.

25. $4 \cdot 16 \overset{?}{=} 2 \cdot 32$

    $64 \overset{?}{=} 64$

    The proportion is true.

27. $9 \cdot 80 \overset{?}{=} 19 \cdot 38$

    $720 \overset{?}{=} 722$

    The proportion is false.

29. $0.5 \cdot 1.3 \overset{?}{=} 1.1 \cdot 0.8$

    $0.65 \overset{?}{=} 0.88$

    The proportion is false.

31. $1.2 \cdot 5.4 \overset{?}{=} 1.8 \cdot 3.6$

    $6.48 \overset{?}{=} 6.48$

    The proportion is true.

33. $1\dfrac{4}{5} \cdot 4\dfrac{1}{6} \overset{?}{=} 2\dfrac{3}{16} \cdot 3\dfrac{3}{7}$

    $\dfrac{9}{5} \cdot \dfrac{25}{6} \overset{?}{=} \dfrac{35}{16} \cdot \dfrac{24}{7}$

    $\dfrac{\cancel{3} \cdot 3 \cdot \cancel{5} \cdot 5}{\cancel{5} \cdot 2 \cdot \cancel{3}} \overset{?}{=} \dfrac{5 \cdot \cancel{7} \cdot \cancel{4} \cdot \cancel{2} \cdot 3}{2 \cdot \cancel{2} \cdot \cancel{4} \cdot \cancel{7}}$

    $\dfrac{15}{2} \overset{?}{=} \dfrac{15}{2}$

    The proportion is true.

35. $\dfrac{1}{5} \cdot 11\dfrac{2}{3} \overset{?}{=} 1\dfrac{1}{7} \cdot 1\dfrac{1}{6}$

    $\dfrac{1}{5} \cdot \dfrac{35}{3} \overset{?}{=} \dfrac{8}{7} \cdot \dfrac{7}{6}$

    $\dfrac{\cancel{5} \cdot 7}{\cancel{5} \cdot 3} \overset{?}{=} \dfrac{\cancel{2} \cdot 4 \cdot \cancel{7}}{\cancel{7} \cdot \cancel{2} \cdot 3}$

    $\dfrac{7}{3} \overset{?}{=} \dfrac{4}{3}$

    The proportion is false.

37. $\dfrac{18}{54} \overset{?}{=} \dfrac{3}{9}$

    $18 \cdot 9 \overset{?}{=} 3 \cdot 54$

    $162 \overset{?}{=} 162$

    Yes the numbers are proportional.

39. $\dfrac{8}{6} \overset{?}{=} \dfrac{21}{16}$

    $8 \cdot 16 \overset{?}{=} 21 \cdot 6$

    $128 \overset{?}{=} 126$

    No the numbers are not proportional.

41. $\dfrac{5}{10} = \dfrac{3}{c}$

    $5 \cdot c = 3 \cdot 10$

    $\dfrac{5 \cdot c}{5} = \dfrac{30}{5}$

    $c = 6$

    Check:

    $\dfrac{5}{10} \overset{?}{=} \dfrac{3}{6}$

    $5 \cdot 6 \overset{?}{=} 3 \cdot 10$

    $30 = 30$

43. $\dfrac{2}{3} = \dfrac{x}{6}$

$2 \cdot 6 = x \cdot 3$

$\dfrac{12}{3} = \dfrac{x \cdot 3}{3}$

$4 = x$

Check:

$\dfrac{2}{3} \stackrel{?}{=} \dfrac{4}{6}$

$2 \cdot 6 \stackrel{?}{=} 4 \cdot 3$

$12 = 12$

45. $\dfrac{0.6}{9.6} = \dfrac{x}{4.8}$

$0.6 \cdot 4.8 = x \cdot 9.6$

$\dfrac{2.88}{9.6} = \dfrac{x \cdot 9.6}{9.6}$

$0.3 = x$

Check:

$\dfrac{0.6}{9.6} \stackrel{?}{=} \dfrac{0.3}{4.8}$

$0.6 \cdot 4.8 \stackrel{?}{=} 0.3 \cdot 9.6$

$2.88 = 2.88$

47. $\dfrac{2.75}{x} = \dfrac{1.5}{1.2}$

$2.75 \cdot 1.2 = 1.5 \cdot x$

$\dfrac{3.3}{1.5} = \dfrac{1.5 \cdot x}{1.5}$

$2.2 = x$

Check:

$\dfrac{2.75}{2.2} \stackrel{?}{=} \dfrac{1.5}{1.2}$

$2.75 \cdot 1.2 \stackrel{?}{=} 1.5 \cdot 2.2$

$3.3 = 3.3$

49. $\dfrac{x}{1\frac{1}{2}} = \dfrac{10\frac{1}{2}}{4\frac{1}{2}}$

$\dfrac{x}{\frac{3}{2}} = \dfrac{\frac{21}{2}}{\frac{9}{2}}$

$x \cdot \dfrac{9}{2} = \dfrac{21}{2} \cdot \dfrac{3}{2}$

$\dfrac{x \cdot \frac{9}{2}}{\frac{9}{2}} = \dfrac{\frac{21}{2} \cdot \frac{3}{2}}{\frac{9}{2}}$

$x = \dfrac{21}{2} \cdot \dfrac{3}{2} \cdot \dfrac{2}{9}$

$x = \dfrac{\cancel{3} \cdot 7 \cdot \cancel{3} \cdot \cancel{2}}{\cancel{2} \cdot 2 \cdot \cancel{3} \cdot \cancel{3}}$

$x = \dfrac{7}{2} = 3\dfrac{1}{2}$

Check:

$\dfrac{3\frac{1}{2}}{1\frac{1}{2}} \stackrel{?}{=} \dfrac{10\frac{1}{2}}{4\frac{1}{2}}$

$\dfrac{\frac{7}{2}}{\frac{3}{2}} \stackrel{?}{=} \dfrac{\frac{21}{2}}{\frac{9}{2}}$

$\dfrac{7}{2} \cdot \dfrac{9}{2} \stackrel{?}{=} \dfrac{21}{2} \cdot \dfrac{3}{2}$

$\dfrac{63}{4} = \dfrac{63}{4}$

51.

$$\frac{x}{1\frac{1}{6}} = \frac{2\frac{5}{8}}{3\frac{1}{2}}$$

$$\frac{x}{\frac{7}{6}} = \frac{\frac{21}{8}}{\frac{7}{2}}$$

$$\frac{7}{2} \cdot x = \frac{21}{8} \cdot \frac{7}{6}$$

$$\frac{\frac{7}{2} \cdot x}{\frac{7}{2}} = \frac{\frac{21}{8} \cdot \frac{7}{6}}{\frac{7}{2}}$$

$$x = \frac{21}{8} \cdot \frac{7}{6} \cdot \frac{2}{7}$$

$$x = \frac{\cancel{3} \cdot \cancel{7} \cdot 7 \cdot \cancel{2}}{2 \cdot 2 \cdot 2 \cdot \cancel{2} \cdot \cancel{3} \cdot \cancel{7}}$$

$$x = \frac{7}{8}$$

Check:

$$\frac{\frac{7}{8}}{1\frac{1}{6}} \stackrel{?}{=} \frac{2\frac{5}{8}}{3\frac{1}{2}}$$

$$\frac{\frac{7}{8}}{\frac{7}{6}} \stackrel{?}{=} \frac{\frac{21}{8}}{\frac{7}{2}}$$

$$\frac{7}{8} \cdot \frac{7}{2} \stackrel{?}{=} \frac{21}{8} \cdot \frac{7}{6}$$

$$\frac{49}{16} = \frac{49}{16}$$

TRY IT YOURSELF

53.

$$\frac{4{,}000}{x} = \frac{3.2}{2.8}$$

$$4{,}000 \cdot 2.8 = 3.2 \cdot x$$

$$11{,}200 = 3.2 \cdot x$$

$$3{,}500 = x$$

55.

$$\frac{12}{6} = \frac{x}{\frac{1}{4}}$$

$$12 \cdot \frac{1}{4} = x \cdot 6$$

$$3 = x \cdot 6$$

$$\frac{3}{6} = x$$

$$\frac{1}{2} = x$$

57.

$$\frac{x}{800} = \frac{900}{200}$$

$$x \cdot 200 = 900 \cdot 800$$

$$x \cdot 200 = 720{,}000$$

$$x = 3{,}600$$

59.

$$\frac{x}{2.5} = \frac{3.7}{9.25}$$

$$x \cdot 9.25 = 3.7 \cdot 2.5$$

$$x \cdot 9.25 = 9.25$$

$$x = 1$$

61.

$$\frac{0.8}{2} = \frac{x}{5}$$

$$0.8 \cdot 5 = 2 \cdot x$$

$$4 = 2 \cdot x$$

$$2 = x$$

63.

$$\frac{x}{4\frac{1}{10}} = \frac{3\frac{3}{4}}{1\frac{7}{8}}$$

$$\frac{x}{\frac{41}{10}} = \frac{\frac{15}{4}}{\frac{15}{8}}$$

$$x \cdot \frac{15}{8} = \frac{15}{4} \cdot \frac{41}{10}$$

$$x = \frac{15}{4} \cdot \frac{41}{10} \cdot \frac{8}{15}$$

$$x = \frac{41}{5} = 8\frac{1}{5}$$

Chapter 5 Ratio, Proportion, and Measurement 117

65. $\dfrac{340}{51} = \dfrac{x}{27}$

$340 \cdot 27 = x \cdot 51$

$9{,}180 = 51 \cdot x$

$180 = x$

67. $\dfrac{0.4}{1.2} = \dfrac{6}{x}$

$0.4 \cdot x = 6 \cdot 1.2$

$0.4 \cdot x = 7.2$

$x = 18$

69. $\dfrac{4.65}{7.8} = \dfrac{x}{5.2}$

$4.65 \cdot 5.2 = x \cdot 7.8$

$24.18 = x \cdot 7.8$

$3.1 = x$

71. $\dfrac{\frac{3}{4}}{\frac{1}{2}} = \dfrac{\frac{1}{4}}{x}$

$\dfrac{3}{4} \cdot x = \dfrac{1}{4} \cdot \dfrac{1}{2}$

$x = \dfrac{1}{4} \cdot \dfrac{1}{2} \cdot \dfrac{4}{3}$

$x = \dfrac{1}{6}$

APPLICATIONS

73. $\dfrac{6}{1.75} = \dfrac{750}{x}$

$6 \cdot x = 1.75 \cdot 750$

$6x = 1{,}312.5$

$x = \$218.75$

75. $\dfrac{12}{57.99} = \dfrac{16}{x}$

$12 \cdot x = 57.99 \cdot 16$

$12 \cdot x = 927.84$

$x = \$77.32$

77. $\dfrac{10}{4} \overset{?}{=} \dfrac{25}{10}$

$10 \cdot 10 \overset{?}{=} 25 \cdot 4$

$100 \overset{?}{=} 100$

Yes the ratios are equal.

79. $\dfrac{3}{7} = \dfrac{x}{56}$

$3 \cdot 56 = 7 \cdot x$

$168 = 7 \cdot x$

$x = 24$

81. $\dfrac{5}{25} = \dfrac{195}{x}$

$5 \cdot x = 25 \cdot 195$

$5 \cdot x = 4875$

$x = 975$

83. $\dfrac{280}{7} = \dfrac{x}{2}$

$280 \cdot 2 = 7 \cdot x$

$560 = 7 \cdot x$

$x = 80\,ft.$

85. $\dfrac{87}{1} = \dfrac{x}{.75}$

$1 \cdot x = .75 \cdot 87$

$x = 65.25\,ft.$

(or 65 ft. 3 in.)

87. $\dfrac{160}{1} = \dfrac{420}{x}$

$160 \cdot x = 1 \cdot 420$

$x = \dfrac{420}{160}$

$x = 2.625\,in.$

89.
$$\frac{1\frac{1}{4}}{3\frac{1}{2}} = \frac{x}{12}$$
$$\frac{5}{4} \cdot 12 = \frac{7}{2} \cdot x$$
$$15 = \frac{7}{2} \cdot x$$
$$\frac{2}{7} \cdot 15 = x$$
$$x = 4\frac{2}{7}$$

91.
$$\frac{15}{2.85} = \frac{100}{x}$$
$$15 \cdot x = 2.85 \cdot 100$$
$$15 \cdot x = 285$$
$$x = 19 \sec.$$

93.
$$\frac{5}{6} = \frac{x}{37.5}$$
$$5 \cdot 37.5 = 6 \cdot x$$
$$187.5 = 6 \cdot x$$
$$x = 31.25 in$$

95.
$$\frac{412}{40} = \frac{x}{30}$$
$$412 \cdot 30 = 40 \cdot x$$
$$12,360 = 40 \cdot x$$
$$x = \$309$$

WRITING

97. A ratio is how two quantities relate to one another. A proportion is how two ratios are related.

99. If 4 ounces of cashews have 639 calories, how many calories are in 10 ounces of cashews?

REVIEW

101. $7.4 + 6.78 + 35 + 0.008 = 49.188$

103. $48.8 - 17.372 = 31.428$

105.
$$-3.8 - (-7.9)$$
$$= -3.8 + 7.9$$
$$= 4.1$$

107. $-35.1 - 13.99 = -49.09$

Section 5.3: American Units of Measurement

VOCABULARY

1. A ruler is used for measuring <u>length</u>.

3. $\frac{3ft}{1yd}, \frac{1ton}{2,000lb}$, and $\frac{4qt}{1gal}$ are examples of <u>unit</u> conversion factors.

5. Some examples of American units of <u>capacity</u> are cups, pints, quarts, and gallons.

CONCEPTS

7. a. 12 inches = 1 foot

   b. 3 feet = 1 yard

   c. 1 yard = 36 inches

   d. 1 mile = 5,280 feet

9. a. 1 cup = 8 fluid ounces

   b. 1 pint = 2 cups

   c. 2 pints = 1 quart

   d. 4 quarts = 1 gallon

11. The value of any unit conversion factor is 1.

13. a. oz can be removed.

    b. lb remains

15. a. $\frac{1 ton}{2,000 lb}$

    b. $\frac{2 pt}{1 qt}$

17. a. iv

    b. i

c. ii

d. iii

19. a. iii

b. iv

c. i

d. ii

NOTATION

21. a. pound

b. ounce

c. fluid ounce

23. $2\,yd = \dfrac{2\,yd}{1} \cdot \dfrac{36\,in}{1\,yd}$

$= 2 \cdot 36\,in$

$= 72\,in$

25. $1\,ton = \dfrac{1\,ton}{1} \cdot \dfrac{2{,}000\,lb}{1\,ton} \cdot \dfrac{16\,oz}{1\,lb}$

$= 1 \cdot 2{,}000 \cdot 16\,oz$

$= 32{,}000\,oz$

GUIDED PRACTICE

27. a. 8

b. $\dfrac{5}{8}\,in,\ 1\dfrac{1}{4}\,in,\ 2\dfrac{7}{8}\,in$

29. a. 16

b. $\dfrac{9}{16}\,in,\ 1\dfrac{3}{4}\,in,\ 2\dfrac{3}{16}\,in$

31. $2\dfrac{9}{16}\,in$

33. $10\dfrac{7}{8}\,in$

35. $\dfrac{4\,yd}{1} \cdot \dfrac{3\,ft}{1\,yd} = 12\,ft$

37. $\dfrac{35\,yd}{1} \cdot \dfrac{3\,ft}{1\,yd} = 105\,ft$

39. $3\dfrac{1}{2}\,ft = \dfrac{7}{2}\,ft \cdot \dfrac{12\,in}{1\,ft} = \dfrac{84}{2}\,in = 42\,in$

41. $5\dfrac{1}{4}\,ft = \dfrac{21}{4}\,ft \cdot \dfrac{12\,in}{1\,ft} = \dfrac{252}{4}\,in = 63\,in$

43. $\dfrac{105\,yd}{1} \cdot \dfrac{3\,ft}{1\,yd} \cdot \dfrac{1\,mi}{5{,}280\,ft}$

$= \dfrac{415}{5{,}280}\,mi = \dfrac{21}{352}\,mi$

$\approx 0.06\,mi$

45. $\dfrac{1{,}540\,yd}{1} \cdot \dfrac{3\,ft}{1\,yd} \cdot \dfrac{1\,mi}{5{,}280\,ft}$

$= \dfrac{4{,}620}{5{,}280}\,mi = \dfrac{7}{8}\,mi$

$= 0.875\,mi$

47. $\dfrac{44\,oz}{1} \cdot \dfrac{1\,lb}{16\,oz} = \dfrac{44}{16}\,lb = \dfrac{11}{4}\,lb$

$= 2\dfrac{3}{4}\,lb = 2.75\,lb$

49. $\dfrac{72\,oz}{1} \cdot \dfrac{1\,lb}{16\,oz} = \dfrac{72}{16}\,lb = \dfrac{9}{2}\,lb$

$= 4\dfrac{1}{2}\,lb = 4.5\,lb$

51. $\dfrac{50\,lb}{1} \cdot \dfrac{16\,oz}{1\,lb} = 50 \cdot 16\,oz = 800\,oz$

53. $\dfrac{87\,lb}{1} \cdot \dfrac{16\,oz}{1\,lb} = 87 \cdot 16\,oz = 1{,}392\,oz$

55. $\dfrac{8\,pt}{1} \cdot \dfrac{2\,c}{1\,pt} \cdot \dfrac{8\,fl\,oz}{1\,c} = 8 \cdot 2 \cdot 8\,fl\,oz = 128\,fl\,oz$

57. $\dfrac{21\,pt}{1} \cdot \dfrac{2\,c}{1\,pt} \cdot \dfrac{8\,fl\,oz}{1\,c} = 21 \cdot 2 \cdot 8\,fl\,oz = 336\,fl\,oz$

59. $\dfrac{165\,\text{min}}{1} \cdot \dfrac{1\,hr}{60\,\text{min}} = \dfrac{165}{60}hr$
$= \dfrac{11}{4}hr = 2\dfrac{3}{4}hr$

61. $\dfrac{330\,\text{min}}{1} \cdot \dfrac{1\,hr}{60\,\text{min}} = \dfrac{330}{60}hr$
$= \dfrac{11}{2}hr = 5\dfrac{1}{2}hr$

TRY IT YOURSELF

63. $\dfrac{3\,qt}{1} \cdot \dfrac{2\,pt}{1\,qt} = 6\,qt$

65. $\dfrac{7{,}200\,\text{min}}{1} \cdot \dfrac{1\,hr}{60\,\text{min}} \cdot \dfrac{1\,\text{day}}{24\,hr}$
$= \dfrac{7{,}200}{1{,}440}\,\text{days} = 5\,\text{days}$

67. $\dfrac{56\,in}{1} \cdot \dfrac{1\,ft}{12\,in} = \dfrac{56}{12}ft$
$= \dfrac{14}{3}ft = 4\dfrac{2}{3}ft$

69. $\dfrac{4\,ft}{1} \cdot \dfrac{12\,in}{1\,ft} = 4 \cdot 12\,in = 48\,in$

71. $\dfrac{16\,pt}{1} \cdot \dfrac{1\,qt}{2\,pt} \cdot \dfrac{1\,gal}{4\,qt} = \dfrac{16}{8}gal = 2\,gal$

73. $\dfrac{80\,oz}{1} \cdot \dfrac{1\,lb}{16\,oz} = \dfrac{80}{16}lb = 5\,lb$

75. $\dfrac{240\,\text{min}}{1} \cdot \dfrac{1\,hr}{60\,\text{min}} = \dfrac{240}{60}hr = 4\,hr$

77. $\dfrac{8\,yd}{1} \cdot \dfrac{3\,ft}{1\,yd} \cdot \dfrac{12\,in}{1\,ft} = 8 \cdot 3 \cdot 12\,in = 288\,in$

79. $\dfrac{90\,in}{1} \cdot \dfrac{1\,ft}{12\,in} \cdot \dfrac{1\,yd}{3\,ft} = \dfrac{90}{36}yd = \dfrac{5}{2}yd = 2\dfrac{1}{2}yd$

81. $\dfrac{5\,yd}{1} \cdot \dfrac{3\,ft}{1\,yd} = 5 \cdot 3\,ft = 15\,ft$

83. $\dfrac{12.4\,\text{tons}}{1} \cdot \dfrac{2{,}000\,lb}{1\,\text{ton}} = 12.4 \cdot 2{,}000\,lb$
$= 24{,}800\,lb$

85. $\dfrac{7\,ft}{1} \cdot \dfrac{1\,yd}{3\,ft} = \dfrac{7}{3}yd = 2\dfrac{1}{3}yd$

87. $\dfrac{15{,}840\,ft}{1} \cdot \dfrac{1\,mi}{5{,}280\,ft} = \dfrac{15{,}840}{5{,}280}mi = 3\,mi$

89. $\dfrac{1}{2}mi \cdot \dfrac{5{,}280\,ft}{1\,mi} = \dfrac{5{,}280}{2}ft = 2{,}640\,ft$

91. $\dfrac{7{,}000\,lb}{1} \cdot \dfrac{1\,\text{ton}}{2{,}000\,lb} = \dfrac{7{,}000}{2{,}000}\,\text{tons}$
$= \dfrac{7}{2}\,\text{tons} = 3\dfrac{1}{2}\,\text{tons}$

93. $\dfrac{32\,fl\,oz}{1} \cdot \dfrac{1\,c}{8\,fl\,oz} \cdot \dfrac{1\,pt}{2\,c} = \dfrac{32}{16}pt = 2\,pt$

APPLICATIONS

95. $\dfrac{450\,ft}{1} \cdot \dfrac{1\,yd}{3\,ft} = \dfrac{450}{3}yd = 150\,yd$

97. $\dfrac{240\,ft}{1} \cdot \dfrac{12\,in}{1\,ft} = 240 \cdot 12\,in = 2{,}880\,in$

99. $\dfrac{1{,}454\,ft}{1} \cdot \dfrac{1\,mi}{5{,}280\,ft} = \dfrac{1{,}454}{5{,}280}mi = 0.28\,mi$

101. $\dfrac{35\,mi}{1} \cdot \dfrac{5{,}280\,ft}{1\,mi} \cdot \dfrac{1\,yd}{3\,ft}$
$= \dfrac{35 \cdot 5{,}280}{3}yd$
$= 61{,}600\,yd$

103. $\dfrac{8\,lb}{1} \cdot \dfrac{16\,oz}{1\,lb} = 8 \cdot 16\,oz = 128\,oz$

105. $\dfrac{9{,}900\,lb}{1} \cdot \dfrac{1\,\text{ton}}{2{,}000\,lb} = \dfrac{9{,}900}{2{,}000}\,\text{tons} = 4.95\,\text{tons}$

107. $\dfrac{17\,gal}{1} \cdot \dfrac{4\,qt}{1\,gal} = 17 \cdot 4\,qt = 68\,qt$

109. $\dfrac{575\,pt}{1} \cdot \dfrac{1\,qt}{2\,pt} \cdot \dfrac{1\,gal}{4\,qt} = \dfrac{575}{8}\,gal = 71.875\,gal$

111. $\dfrac{2.5\,gal}{1} \cdot \dfrac{4\,qt}{1\,gal} \cdot \dfrac{2\,pt}{1\,qt} \cdot \dfrac{2\,c}{1\,pt} \cdot \dfrac{16\,oz}{1\,c}$
$= 2.5 \cdot 128\,oz = 320\,oz$

113. $\dfrac{147\,hr}{1} \cdot \dfrac{1\,day}{24\,hr} = \dfrac{147}{24}\,day = 6.125\,days$

WRITING

115. a. Write a form of 1 that uses feet and inches: $\dfrac{1\,ft}{12\,in}$ or $\dfrac{12\,in}{1\,ft}$

b. Write a form of 1 that involves pints and gallons: $\dfrac{8\,pt}{1\,gal}$ or $\dfrac{1\,gal}{8\,pt}$

REVIEW

117. a. 3,700

b. 3,670

c. 3,673.26

d. 3,673.3

Section 5.4: Metric Units of Measurement

VOCABULARY

1. The meter, the gram, and the liter are basic units of measurement in the metric system.

3. a. *Deka* means tens.

b. *Hecto* means hundreds.

c. *Kilo* means thousands.

5. We can convert from one unit to another in the metric system using unit conversion factors or a conversion chart like the one shown below.

7. The weight of an object is determined by the Earth's gravitational pull on the object.

CONCEPTS

9. a. 1 kilometer = 1,000 meters

b. 100 centimeters = 1 meter

c. 1,000 millimeters = 1 meter

11. a. 1,000 milliliters = 1 liter

b. 1 dekaliter = 10 liters

13. a. $\dfrac{1\,km}{1,000\,m}$

b. $\dfrac{100\,cg}{1\,g}$

c. $\dfrac{1,000\text{ millilters}}{1\text{ liter}}$

15. a. iii

b. i

c. ii

17. a. ii

b. iii

c. i

NOTATION

19. $20\,cm = \dfrac{20\,cm}{1} \cdot \dfrac{1\,m}{100\,cm}$
$= \dfrac{20}{100}\,m$
$= .2\,m$

21. $0.2\,kg = \dfrac{0.2\,kg}{1} \cdot \dfrac{1,000\,g}{1\,kg} \cdot \dfrac{1,000\,mg}{1\,g}$
$= 0.2 \cdot 1,000 \cdot 1,000\,mg$
$= 200,000\,mg$

GUIDED PRACTICE

23. $1\,cm$, $3\,cm$, $5\,cm$

25. a. 10, 1 millimeter

b. $27\,mm$, $41\,mm$, $55\,mm$

27. $156 mm$

29. $280 mm$

31. $\dfrac{380cm}{1} \cdot \dfrac{1m}{100cm} = \dfrac{380}{100}m = 3.8m$

33. $\dfrac{120cm}{1} \cdot \dfrac{1m}{100cm} = \dfrac{120}{100}m = 1.2m$

35. $\dfrac{8.7m}{1} \cdot \dfrac{1000mm}{1m} = 8.7 \cdot 1000mm = 8,700mm$

37. $\dfrac{2.89m}{1} \cdot \dfrac{1000mm}{1m} = 2.89 \cdot 1000mm = 2,890 mn$

39. $\dfrac{4.5cm}{1} \cdot \dfrac{1m}{100cm} \cdot \dfrac{1km}{1000m} = \dfrac{4.5}{100,000}km$
    $= 0.000045 km$

41. $\dfrac{0.3cm}{1} \cdot \dfrac{1m}{100cm} \cdot \dfrac{1km}{1000m} = \dfrac{0.3}{100,000}km$
    $= 0.000003 km$

43. $\dfrac{1.93kg}{1} \cdot \dfrac{1,000g}{1kg} = 1.93 \cdot 1,000g = 1,930g$

45. $\dfrac{4.531kg}{1} \cdot \dfrac{1,000g}{1kg} = 4.531 \cdot 1,000g = 4,531g$

47. $\dfrac{6,000mg}{1} \cdot \dfrac{1g}{1,000mg} = \dfrac{6,000}{1,000}g = 6g$

49. $\dfrac{3,500mg}{1} \cdot \dfrac{1g}{1,000mg} = \dfrac{3,500}{1,000}g = 3.5g$

51. $\dfrac{3L}{1} \cdot \dfrac{1,000mL}{1L} = 3 \cdot 1,000mL = 3,000mL$

53. $\dfrac{26.3L}{1} \cdot \dfrac{1,000mL}{1L}$
    $= 26.3 \cdot 1,000mL = 26,300mL$

TRY IT YOURSELF

55. Refer to a conversion chart: 1 place to the right

    $0.31dm = 3.1cm$

57. Refer to a conversion chart: 3 places to the left

    $500mL = 0.5L$

59. Refer to a conversion chart: 3 places to the right

    $2kg = 2,000g$

61. Refer to a conversion chart: 1 place to the right

    $0.074cm = 0.74mm$

63. Refer to a conversion chart: 3 places to the right

    $1,000kg = 1000,000g$

65. Refer to a conversion chart: 3 places to the left

    $658.23L = 0.65823kL$

67. Refer to a conversion chart: 1 place to the left

    $4.72cm = 0.472dm$

69. $10mL = 10cc$

71. Refer to a conversion chart: 3 places to the left

    $500mg = 0.5g$

73. Refer to a conversion chart: 3 places to the left

    $5,689g = 5.689kg$

75. Refer to a conversion chart: 2 places to the left

    $453.2cm = 4.532m$

77. Refer to a conversion chart: 1 place to the left

    $0.325dL = 0.0325L$

79. Refer to a conversion chart: 3 places to the right

    $675dam = 675,000cm$

81. Refer to a conversion chart: 3 places to the left

    $0.00777cm = 0.0000077dam$

83. Refer to a conversion chart: 2 places to the left

    $134m = 1.34hm$

85. Refer to a conversion chart: 2 places to the right

    $65.78km = 6,578dam$

APPLICATIONS

87. $\dfrac{500}{1000} = .5km \qquad \dfrac{1,000}{1,000} = 1km \qquad \dfrac{1,500}{1,000} = 1.5km$

$\dfrac{5,000}{1,000} = 5km \qquad \dfrac{10,000}{1,000} = 10km$

89. $\dfrac{343m}{1} \cdot \dfrac{1hm}{100m} = \dfrac{343}{100}hm = 3.43hm$

91. $\dfrac{120mm}{1} \cdot \dfrac{1cm}{10mm} = \dfrac{120}{10}cm = 12cm$

$\dfrac{80mm}{1} \cdot \dfrac{1cm}{10mm} = \dfrac{80}{10}cm = 8cm$

93. $\dfrac{0.05mL}{1} \cdot \dfrac{1L}{1,000mL} = \dfrac{0.05}{1,000}L = 0.00005L$

95. $60(50) = 3,000mg$

$\dfrac{3,000mg}{1} \cdot \dfrac{1g}{1,000mg} = 3g$

97. $0.5 \cdot 6 = 3L$

$\dfrac{3L}{1} \cdot \dfrac{1,000mL}{1L} = 3,000mL$

99. 3 bottles would only have 0.852kg – you would need 4 bottles.

101. $3cc = 3mL$

WRITING

103. We move 3 places to the right because we are multiplying by 1,000.

105. century – 100 years

centipede – 100 legs (an exaggeration)

centennial – 100th birthday

cent – 1/100 of a dollar

centiliter – 1/100 of a liter

REVIEW

107.
$$\begin{array}{r} 0.88\ldots \\ 9\overline{)8.00} \\ \underline{-72} \\ 80 \\ \underline{-72} \\ 8 \end{array}$$

$0.\overline{8}$

109.
$$\begin{array}{r} 0.077\cdots \\ 90\overline{)7.000} \\ \underline{-630} \\ 700 \\ \underline{-630} \\ 700 \end{array}$$

$\dfrac{7}{90} = .0777\ldots = 0.0\overline{7}$

Section 5.5: Converting between American and Metric Units

VOCABULARY

1. In the American system, temperatures are measured in degrees <u>Fahrenheit</u>. In the metric system, temperatures are measured in degrees <u>Celsius</u>.

CONCEPTS

3. a. meter

b. meter

c. inch

d. mile

5. a. liter

b. liter

c. gallon

7. a. $\dfrac{0.03m}{1ft}$

b. $\dfrac{0.45kg}{1lb}$

c. $\dfrac{3.79L}{1gal}$

NOTATION

9. $4,500ft \approx \dfrac{4,500ft}{1} \cdot \dfrac{0.30m}{1ft}$

$\approx 1,350m$

11. $3kg \approx \dfrac{3kg}{1} \cdot \dfrac{1,000g}{1kg} \cdot \dfrac{0.035oz}{1g}$

$\approx 3 \cdot 1,000 \cdot 0.035oz$

$\approx 105oz$

GUIDED PRACTICE

13. $\dfrac{25cm}{1} \cdot \dfrac{0.39in}{1cm} \approx 9.75in \approx 10in$

15. $\dfrac{88cm}{1} \cdot \dfrac{0.39in}{1cm} \approx 34.32in \approx 34in$

17. $\dfrac{8,400ft}{1} \cdot \dfrac{0.3m}{1ft} \approx 2,520m$

19. $\dfrac{25,115ft}{1} \cdot \dfrac{0.3m}{1ft} \approx 7,534.5m$

21. $\dfrac{20lb}{1} \cdot \dfrac{16oz}{1lb} \cdot \dfrac{28.35g}{1oz} \approx 9,072g$

23. $\dfrac{75lb}{1} \cdot \dfrac{16oz}{1lb} \cdot \dfrac{28.35g}{1oz} \approx 34,020g$

25. $\dfrac{6.5kg}{1} \cdot \dfrac{2.2lb}{1kg} \approx 14.3lb$

27. $\dfrac{300kg}{1} \cdot \dfrac{2.2lb}{1kg} \approx 660lb$

29. $\dfrac{650mL}{1} \cdot \dfrac{1L}{1,000mL} \cdot \dfrac{1.06qt}{1L}$

$\approx \dfrac{689}{1000}qt \approx .689qt \approx 0.7qt$

31. $\dfrac{1,200mL}{1} \cdot \dfrac{1L}{1,000mL} \cdot \dfrac{1.06qt}{1L}$

$\approx \dfrac{1,272}{1000}qt \approx 1.272qt \approx 1.3qt$

33. $C = \dfrac{5}{9}(120 - 32)$

$= \dfrac{5}{9}(88) = \dfrac{440}{9}$

$\approx 48.8888... \approx 48.9°C$

35. $C = \dfrac{5}{9}(35 - 32)$

$= \dfrac{5}{9}(3) = \dfrac{15}{9}$

$\approx 1.6666... \approx 1.7°C$

37. $F = \dfrac{9}{5}(75) + 32$

$= 135 + 32 = 167°F$

39. $F = \dfrac{9}{5}(10) + 32$

$= 18 + 32 = 40°F$

TRY IT YOURSELF

41. $\dfrac{25lb}{1} \cdot \dfrac{16oz}{1lb} \cdot \dfrac{28.35g}{1oz} \approx 11,340g$

43. $F = \dfrac{9}{5}(50) + 32$

$= 90 + 32 = 122°F$

45. $\dfrac{0.75qt}{1} \cdot \dfrac{1L}{1.06qt} \cdot \dfrac{1,000mL}{1L}$

$\approx \dfrac{750}{1.06}mL \approx 707.5mL$

47. $\dfrac{0.5kg}{1} \cdot \dfrac{1lb}{0.45kg} \cdot \dfrac{16oz}{1lb}$

$\approx \dfrac{8}{0.45} oz \approx 17.8 oz$

49. $\dfrac{3.75m}{1} \cdot \dfrac{1yd}{0.91m} \cdot \dfrac{3ft}{1yd} \cdot \dfrac{12in}{1ft}$

$\approx \dfrac{135}{0.91} in \approx 148.4 in$

51. $\dfrac{3 fl\, oz}{1} \cdot \dfrac{1L}{33.8 fl\, oz} \approx 0.1L$

53. $\dfrac{12km}{1} \cdot \dfrac{1mi}{1.61km} \cdot \dfrac{5,280ft}{1mi}$

$\approx \dfrac{63,360}{1.61} ft \approx 39,354 ft$

55. $\dfrac{37oz}{1} \cdot \dfrac{28.35g}{1oz} \cdot \dfrac{1kg}{1,000g}$

$\approx \dfrac{1,048.95}{1,000} kg \approx 1.0 kg$

57. $F = \dfrac{9}{5}(-10) + 32$

$= -18 + 32 = 14°F$

59. $\dfrac{17g}{1} \cdot \dfrac{0.035oz}{1g} \approx .6oz$

61. $\dfrac{7.2L}{1} \cdot \dfrac{33.81 fl\, oz}{1L} \approx 243.4 fl\, oz$

63. $\dfrac{3ft}{1} \cdot \dfrac{12in}{1ft} \cdot \dfrac{2.54cm}{1in} \approx 91.4 cm$

65. $\dfrac{500mL}{1} \cdot \dfrac{1L}{1,000mL} \cdot \dfrac{1.06qt}{1L}$

$\approx \dfrac{530}{1,000} qt \approx .53qt \approx .5qt$

67. $C = \dfrac{5}{9}(50 - 32)$

$= \dfrac{5}{9}(18) = 10°C$

69. $\dfrac{5,000in}{1} \cdot \dfrac{1ft}{12in} \cdot \dfrac{1m}{3.28ft}$

$\approx \dfrac{5,000}{39.36} m \approx 127 m$

71. $C = \dfrac{5}{9}(-5 - 32)$

$= \dfrac{5}{9}(-37) = \dfrac{-185}{9}$

$\approx -20.5555... \approx -20.6°C$

APPLICATIONS

73. $\dfrac{8km}{1} \cdot \dfrac{.62mi}{1km} = 4.96 mi$

About 5 miles.

75. $\dfrac{112km}{1} \cdot \dfrac{.62mi}{1km} = 69.44 mi$

About 70 mph

77. $\dfrac{6,288ft}{1} \cdot \dfrac{1mi}{5,280ft} \cdot \dfrac{1.61km}{1mi}$

$\approx 1.9 km$

79. $\dfrac{.75in}{1} \cdot \dfrac{2.54cm}{1in} \approx 1.9 cm$

81. $\dfrac{187kg}{1} \cdot \dfrac{2.2lb}{1kg} \approx 411 lb$

$\dfrac{350kg}{1} \cdot \dfrac{2.2lb}{1kg} \approx 770 lb$

83. $\dfrac{8oz}{1} \cdot \dfrac{28.35g}{1oz} \approx 226.8 g$

$\dfrac{8 fl\, oz}{1} \cdot \dfrac{29.57mL}{1 fl\, oz} \cdot \dfrac{1L}{1,000mL} \approx 0.24L$

85. $\dfrac{32kg}{1} \cdot \dfrac{2.2lb}{1kg} \approx 70.4 lb$, so no

87.
$$C = \frac{5}{9}(143-32)$$
$$= \frac{5}{9}(111)$$
$$= 61.\overline{6}$$

About 62°C

89. 15°C = 59°F – too cold.

50°C = 122°F – too hot.

I'd choose the 20°C (approximately 82° F.)

91. -5°C and 0°C only: 10°C is too warm for snow.

93. The three quart purchase: $\frac{\$4.50}{3qt} = \$1.50/qt$

The two liter purchase: $\frac{\$3.6}{2L} \cdot \frac{1L}{1.06} \approx \$1.70/qt$

You should go with the 3 quart purchase.

WRITING

95. Convert to miles by multiplying the number of kilometers 0.62 or by dividing by 1.61.

97. It is possibly costing money because people can't easily compare the prices to see if the United States prices are a better deal.

REVIEW

99. $\frac{3}{5} + \frac{4}{3} = \frac{9}{15} + \frac{20}{15} = \frac{29}{15}$

101. $\frac{3}{5} \cdot \frac{4}{3} = \frac{4}{5}$

103.
$$\overset{1}{3}.25$$
$$+4.80$$
$$\overline{8.05}$$

105.
$$325$$
$$\times 48$$
$$\overline{2,600}$$
$$+13,000$$
$$\overline{15,600}$$

Since the original factors had a total of 3 places right of the decimal, this becomes 15.6.

Chapter 5 Review

1. $\frac{7}{25}$

3. $\frac{24}{36} = \frac{2 \cdot \cancel{12}}{3 \cdot \cancel{12}} = \frac{2}{3}$

5. $\frac{4in}{12in} = \frac{\cancel{4}\,in}{3 \cdot \cancel{4}\,in} = \frac{1}{3}$

7. $\frac{0.28}{0.35} = \frac{28}{35} = \frac{4 \cdot \cancel{7}}{5 \cdot \cancel{7}} = \frac{4}{5}$

9. $\frac{2\frac{1}{3}}{2\frac{2}{3}} = \frac{\frac{7}{3}}{\frac{8}{3}} = \frac{7}{\cancel{3}} \cdot \frac{\cancel{3}}{8} = \frac{7}{8}$

11. $\frac{15\,min}{3\,hr} = \frac{15\,min}{180\,min} = \frac{\cancel{15}\,min}{12 \cdot \cancel{15}\,min} = \frac{1}{12}$

13. $\frac{64cm}{12yr} = \frac{\cancel{4} \cdot 16cm}{3 \cdot \cancel{4}\,yr} = \frac{16cm}{3yr}$

15. $\frac{600t}{20\,min} = \frac{\cancel{20} \cdot 30t}{\cancel{20}\,min} = 30$ tickets per minute

17. $\frac{195\,ft}{6r} = 32.5$ feet per roll

19. $\frac{\$11.45}{5p} = \$2.29$ per pair

21. $\frac{185}{160} = \frac{\cancel{5} \cdot 37}{\cancel{5} \cdot 32} = \frac{37}{32}$

23. $\dfrac{54,000 \text{ people}}{48 \text{ min}} = 1,125 \text{ people/min}$

1,125 people per minute

25. a. $\dfrac{20}{30} = \dfrac{2}{3}$

b. $\dfrac{6 \text{ buses}}{100 \text{ cars}} = \dfrac{36 \text{ buses}}{600 \text{ cars}}$

27. $\dfrac{8}{12} = \dfrac{2}{3} \neq \dfrac{3}{7}$

The proportion is false.

29. $9 \cdot 6 \stackrel{?}{=} 2 \cdot 27$

$54 \stackrel{?}{=} 54$

The proportion is true.

31. $3.5 \cdot 3 \stackrel{?}{=} 1.2 \cdot 9.3$

$10.5 \stackrel{?}{=} 11.16$

The proportion is false.

33. $\dfrac{20}{36} = \dfrac{\cancel{4} \cdot 5}{\cancel{4} \cdot 9} = \dfrac{5}{9}$

The numbers are proportional.

35. $\dfrac{12}{18} = \dfrac{3}{x}$

$12x = 3 \cdot 18$

$x = \dfrac{54}{12}$

$x = 4.5$

37. $\dfrac{4.8}{6.6} = \dfrac{x}{9.9}$

$4.8 \cdot 9.9 = x \cdot 6.6$

$\dfrac{47.52}{6.6} = x$

$7.2 = x$

39. $\dfrac{1\tfrac{9}{11}}{x} = \dfrac{3\tfrac{1}{3}}{2\tfrac{3}{4}}$

$\left(\dfrac{20}{11}\right)\left(\dfrac{11}{4}\right) = \left(\dfrac{10}{3}\right)x$

$5 = \dfrac{10}{3}x$

$\dfrac{3}{10} \cdot 5 = \dfrac{3}{10} \cdot \dfrac{10}{3}x$

$\dfrac{3}{2} = x = 1\dfrac{1}{2}$

41. $\dfrac{\tfrac{2}{3}}{\tfrac{1}{2}} = \dfrac{x}{0.25}$

$\dfrac{2}{3} \cdot \dfrac{1}{4} = x \cdot \dfrac{1}{2}$

$\dfrac{1}{6} = x \cdot \dfrac{1}{2}$

$2 \cdot \dfrac{1}{6} = x \cdot \dfrac{1}{2} \cdot 2$

$\dfrac{1}{3} = x$

43. $\dfrac{35}{2} = \dfrac{x}{11}$

$35 \cdot 11 = x \cdot 2$

$\dfrac{385}{2} = x$

$192.5 \text{ mi} = x$

45. $\dfrac{\tfrac{1}{8}}{1} = \dfrac{1\tfrac{1}{2}}{x}$

$\dfrac{1}{8}x = 1 \cdot 1\dfrac{1}{2}$

$8 \cdot \dfrac{1}{8}x = 8 \cdot \dfrac{3}{2}$

$x = 12 \text{ ft}$

47. a. 16

b. $\frac{7}{16}in$, $1\frac{1}{2}in$, $1\frac{3}{4}in$, $2\frac{5}{8}in$

49. $\frac{1mi}{5,280ft}=1$; $\frac{5,280ft}{1mi}=1$

51. $\frac{5yd}{1}\cdot\frac{3ft}{1yd}=15ft$

53. $\frac{66in}{1}\cdot\frac{1ft}{12in}=\frac{11}{2}ft=5\frac{1}{2}ft$

55. $\frac{4\frac{1}{2}ft}{1}\cdot\frac{12in}{1ft}=\frac{9}{2}\cdot 12in=54in$

57. $\frac{32oz}{1}\cdot\frac{1lb}{16oz}=2lb$

59. $\frac{3tons}{1}\cdot\frac{2,000lb}{1ton}\cdot\frac{16oz}{1lb}=96,000oz$

61. $\frac{5pt}{1}\cdot\frac{2c}{1pt}\cdot\frac{8fl\,oz}{1c}=80fl\,oz$

63. $\frac{17qt}{1}\cdot\frac{2pt}{1qt}\cdot\frac{2c}{1pt}=68c$

65. $\frac{5gal}{1}\cdot\frac{4qt}{1gal}\cdot\frac{2pt}{1qt}=40pt$

67. $\frac{20\min}{1}\cdot\frac{60\sec}{1\min}=1,200\sec$

69. $\frac{200hr}{1}\cdot\frac{1d}{24hr}=\frac{25}{3}d=8\frac{1}{3}d$

71. $\frac{4.5d}{1}\cdot\frac{24hr}{1d}=108hr$

73. $\frac{210yd}{1}\cdot\frac{3ft}{1yd}\cdot\frac{1mi}{5,280ft}=\frac{21}{176}mi$
    $\approx 0.12mi$

75. $\frac{1,454ft}{1}\cdot\frac{1yd}{3ft}=\frac{1,454}{3}yd=484\frac{2}{3}yd$

77. a. 10, 1mm

b. 19mm, 3cm, 45mm, 62mm

79. a. $\frac{1km}{1,000m}=1$; $\frac{1,000m}{1km}=1$

b. $\frac{1g}{100cg}=1$; $\frac{100cg}{1g}=1$

81. Refer to a conversion chart: 2 places left

$475cm = 4.75m$

83. Refer to a conversion chart: 3 places right

$165.7km = 165,700m$

85. Refer to a conversion chart: 5 places left

$5,000cg = 0.05kg$

87. Refer to a conversion chart: 3 places left

$5,425g = 5.425kg$

89. Refer to a conversion chart: 2 places left

$150cL = 1.5L$

91. Refer to a conversion chart: 1 place left

$400mL = 40cL$

93. $\frac{1,350g}{1}\cdot\frac{1kg}{1,000g}=\frac{1,350}{1,000}kg=1.35kg$

95. $100\cdot 500mg = 50,000mg$
    $\frac{50,000mg}{1}\cdot\frac{1g}{1,000mg}=\frac{50,000}{1,000}g=50g$

97. $\frac{50m}{1}\cdot\frac{3.28ft}{1m}\approx 164ft$

99. $\frac{1,930mi}{1}\cdot\frac{1.61km}{mi}\approx 3,107km$

101. $\dfrac{30oz}{1} \cdot \dfrac{28.35g}{1oz} \approx 850.5g$

103. $\dfrac{50lb}{1} \cdot \dfrac{0.45kg}{1lb} \cdot \dfrac{1,000g}{1kg} \approx 22,500g$

105. $\dfrac{910g}{1} \cdot \dfrac{1kg}{1,000g} \cdot \dfrac{2.2lb}{1kg} \approx \dfrac{2002}{1,000}lb \approx 2lb$

107. $\dfrac{42gal}{1barrel} \cdot \dfrac{4qt}{1gal} \cdot \dfrac{0.95L}{1qt} \approx 159.6L / barrel$

109. $C = \dfrac{5}{9}(77-32) = \dfrac{5}{9}(45)$
$= \dfrac{225}{9} = 25°C$

Chapter 5 Test

1. a. A <u>ratio</u> is the quotient of two numbers or the quotient of two quantities that have the same units.

   b. A <u>rate</u> is the quotient of two quantities that have different units.

   c. A <u>proportion</u> is a statement that two ratios (or rates) are equal.

   d. The <u>cross</u> products for the proportion $\dfrac{3}{8} = \dfrac{6}{16}$ are $3 \cdot 16$ and $8 \cdot 6$.

   e. *Deci* means <u>tenths</u>, *centi* means <u>hundredths</u>, and *milli* means <u>thousandths</u>.

   f. The meter, the gram, and the liter are basic units of measurement in the <u>metric</u> system.

   g. In the American system, temperatures are measured in degrees <u>Fahrenheit</u>. In the metric system, temperatures are measured in degrees <u>Celsius</u>.

3. $\dfrac{6ft}{8ft} = \dfrac{\cancel{2} \cdot 3 \cancel{ft}}{\cancel{2} \cdot 4 \cancel{ft}} = \dfrac{3}{4}$

5. $\dfrac{0.26}{0.65} = \dfrac{26}{65} = \dfrac{2 \cdot \cancel{13}}{5 \cdot \cancel{13}} = \dfrac{2}{5}$

7. $\dfrac{54ft}{36\sec} = \dfrac{3 \cdot 18ft}{2 \cdot 18\sec} = \dfrac{3ft}{2\sec}$

9. $\dfrac{675kwh}{30d} = 22.5$ kilowatt hours per day

11. a. $25 \cdot 3 \stackrel{?}{=} 33 \cdot 2$
$75 \stackrel{?}{=} 66$

   Not a proportion

   b. $2.2 \cdot 2.8 \stackrel{?}{=} 3.5 \cdot 1.76$
$6.16 \stackrel{?}{=} 6.16$

   Is a proportion

13. $\dfrac{x}{3} = \dfrac{35}{7}$
$7x = 35 \cdot 3$
$x = \dfrac{35 \cdot 3}{7} = \dfrac{105}{7}$
$x = 15$

15. $\dfrac{2\frac{2}{9}}{\frac{4}{3}} = \dfrac{x}{1\frac{1}{2}}$

$\left(\dfrac{20}{9}\right)\left(\dfrac{3}{2}\right) = x\left(\dfrac{4}{3}\right)$

$\left(\dfrac{3}{4}\right)\left(\dfrac{20}{9}\right)\left(\dfrac{3}{2}\right) = x$

$x = \dfrac{5}{2} = 2\dfrac{1}{2}$

17. $\dfrac{13oz}{\$2.79} = \dfrac{16oz}{x}$
$13x = 2.79 \cdot 16$
$x = \dfrac{44.64}{13}$
$x \approx \$3.43$

19. a. 16

    b. $\dfrac{5}{16}in$, $1\dfrac{3}{8}in$, $2\dfrac{3}{4}in$

21. $\dfrac{180 in}{1} \cdot \dfrac{1 ft}{12 in} = 15 ft$

23. $\dfrac{10\tfrac{3}{4} lb}{1} \cdot \dfrac{16 oz}{1 lb} = \dfrac{43}{4} \cdot 16 oz = 172 oz$

25. $\dfrac{1 gal}{1} \cdot \dfrac{4 qt}{1 gal} \cdot \dfrac{2 pt}{1 qt} \cdot \dfrac{2 c}{1 pt} \cdot \dfrac{8 fl\,oz}{1 c} = 128\, fl\,oz$

27. a. the one on the left

    b. the longer one

    c. the right side

29. $\dfrac{500 m}{1} \cdot \dfrac{1 km}{1,000 m} = \dfrac{1}{2} km = 0.5 km$

31. $\dfrac{8,000 cg}{1} \cdot \dfrac{1 g}{100 cg} \cdot \dfrac{1 kg}{1000 g} = 0.08 kg$

33. $\dfrac{50 t}{1} \cdot \dfrac{150 mg}{1 t} \cdot \dfrac{1 g}{1,000 mg} = 7.5 g$

35. $\dfrac{160 lb}{1} \cdot \dfrac{0.45 kg}{1 lb} \approx 72 kg$

    Jim weighs more than Ricardo does.

37. $\dfrac{16.5 in}{1} \cdot \dfrac{2.54 cm}{1 in} \approx 41.91 cm \approx 42 cm$

39. A scale is a ratio comparing the size of a drawing to the size of the actual object. For example, a map could be on the scale 1:6, where 1 inch represents 6 miles.

Chapters 1 – 5 Cumulative Review

1. a. five million, seven hundred sixty-four thousand, five hundred two

   b. 5,000,000 + 700,000 + 60,000 + 4,000 + 500 + 2

3.
    70,006
   $-$348
   —————
   69,658

5.
   $37 \overline{)743}$ — quotient 20
   -74
   03
   -0
   3
   20 R 3

7. 1, 2, 3, 5, 6, 10, 15, 30

9. $20 = 2 \cdot 2 \cdot 5$
   $28 = 2 \cdot 2 \cdot 7$
   $GCF(20, 28) = 2 \cdot 2 = 4$
   $LCM(20, 28) = 2 \cdot 2 \cdot 5 \cdot 7 = 140$

11. $-(-10) \,\square\, |-11|$
    $10 \,\square\, 11$
    $-(-10) \,\boxed{<}\, |-11|$

13. $3 - (-12) = 3 + 12 = 15$ shots

15. a. -8

    b. undefined

    c. -8

    d. 0

    e. 8

    f. 0

17. $-3,900 + (-5,800) + 4,700$
    $= -9,700 + 4,700 = -5,000$

19. $\dfrac{9}{10} \cdot \dfrac{6}{6} = \dfrac{54}{60}$

21. $A = \dfrac{1}{2} bh$

23. $\dfrac{11}{12} - \dfrac{7}{15} = \dfrac{11}{12} \cdot \dfrac{5}{5} - \dfrac{7}{15} \cdot \dfrac{4}{4}$
    $= \dfrac{55}{60} - \dfrac{28}{60} = \dfrac{27}{60} = \dfrac{9}{20}$

25. $\dfrac{5}{32}+\dfrac{5}{16}+\dfrac{1}{2}=\dfrac{5}{32}+\dfrac{5}{16}\cdot\dfrac{2}{2}+\dfrac{1}{2}\cdot\dfrac{16}{16}$

$=\dfrac{5}{32}+\dfrac{10}{32}+\dfrac{16}{32}=\dfrac{31}{32}in$

27. $1\dfrac{1}{2}-\dfrac{3}{4}=\dfrac{3}{2}-\dfrac{3}{4}=\dfrac{6}{4}-\dfrac{3}{4}=\dfrac{3}{4}$ hp

29. $-64.22 > -64.238$

31. Subtract the larger absolute value from the smaller one. Since the number with the larger absolute value is negative, the answer will be negative.

$20.04 - 2.4 = 17.64$
$-20.04 + 2.4 = -17.64$

33. Since there are 2 zeros in 100, move the decimal place 2 units right.

$2.5 \cdot 100 = 250$

35. Since there are 2 zeros in 100, move the decimal place 2 units left.

$2.5 \div 100 = 0.025$

37.
$$\begin{array}{r}0.0833\\12\overline{)1.0000}\\-96\\\hline 40\\-36\\\hline 40\\-36\\\hline 4\end{array}$$

$\dfrac{1}{12} = 0.08\overline{3}$

39. $3\sqrt{25} + 4\sqrt{4} = 3 \cdot 5 + 4 \cdot 2$
$= 15 + 8 = 23$

41. 94 pound bag: $\dfrac{\$4.48}{94 lb} = \$0.0477/lb$

100 pound bag: $\dfrac{\$4.80}{100 lb} = \$0.048/lb$

The 94 pound bag is a better deal.

43. $\dfrac{55}{12} = \dfrac{x}{44}$

$55 \cdot 44 = 12 \cdot x$

$\dfrac{2420}{12} = x$

$202 mg \approx x$

45.
a. $\dfrac{40d}{1} \cdot \dfrac{24hr}{1d} = 960 hr$

b. $\dfrac{3d}{1} \cdot \dfrac{24h}{1d} \cdot \dfrac{60\min}{1h} = 4,320\min$

c. $\dfrac{8\min}{1} \cdot \dfrac{60\sec}{1\min} = 480\sec$

47. $\dfrac{2.4m}{1} \cdot \dfrac{1,000mm}{1m} = 2,400mm$

49.
a. $\dfrac{2L}{1} \cdot \dfrac{1.06qt}{L} \approx 2.12qt$

$1 gal = 4 qt$

A 1-gallon bottle is bigger.

b. a meterstick

CHAPTER 6: Percent

Section 6.1: Percents, Decimals, and Fractions

VOCABULARY

1. <u>Percent</u> means parts per one hundred.

CONCEPTS

3. To write a percent as a fraction, drop the % symbol and write the given number over <u>100</u>. Then <u>simplify</u> the fraction, if possible.

5. To write a decimal as a percent, multiply the decimal by 100 by moving the decimal point 2 places to the <u>right</u> and then insert a % symbol.

NOTATION

7. percent

GUIDED PRACTICE

9. 84% shaded, 16% not shaded

11. 107%

13. $\dfrac{99}{100} = 99\%$

15. a. $\dfrac{15}{100} = 15\%$

    b. $100 - 15 = 85\%$

17. $\dfrac{17}{100}$

19. $\dfrac{91}{100}$

21. $\dfrac{4}{100} = \dfrac{\cancel{4}}{\cancel{4} \cdot 25} = \dfrac{1}{25}$

23. $\dfrac{60}{100} = \dfrac{3 \cdot \cancel{20}}{5 \cdot \cancel{20}} = \dfrac{3}{5}$

25. $1.9\% = \dfrac{1.9}{100} = \dfrac{1.9}{100} \cdot \dfrac{10}{10} = \dfrac{19}{1,000}$

27. $54.7\% = \dfrac{54.7}{100} = \dfrac{54.7}{100} \cdot \dfrac{10}{10} = \dfrac{547}{1,000}$

29. $12.5\% = \dfrac{12.5}{100} = \dfrac{12.5}{100} \cdot \dfrac{10}{10} = \dfrac{125}{1,000}$

    $= \dfrac{\cancel{125}}{8 \cdot \cancel{125}} = \dfrac{1}{8}$

31. $6.8\% = \dfrac{6.8}{100} = \dfrac{6.8}{100} \cdot \dfrac{10}{10} = \dfrac{68}{1,000}$

    $= \dfrac{\cancel{4} \cdot 17}{\cancel{4} \cdot 250} = \dfrac{17}{250}$

33. $1\dfrac{1}{3}\% = \dfrac{\frac{4}{3}}{100} = \dfrac{4}{3} \cdot \dfrac{1}{100}$

    $= \dfrac{\cancel{4}}{3} \cdot \dfrac{1}{\cancel{4} \cdot 25} = \dfrac{1}{75}$

35. $14\dfrac{1}{6}\% = \dfrac{\frac{85}{6}}{100} = \dfrac{85}{6} \cdot \dfrac{1}{100}$

    $= \dfrac{\cancel{5} \cdot 17}{6} \cdot \dfrac{1}{\cancel{5} \cdot 20} = \dfrac{17}{120}$

37. $130\% = \dfrac{130}{100} = \dfrac{13 \cdot \cancel{10}}{10 \cdot \cancel{10}} = \dfrac{13}{10}$

39. $220\% = \dfrac{220}{100} = \dfrac{11 \cdot \cancel{20}}{5 \cdot \cancel{20}} = \dfrac{11}{5}$

41. $0.35\% = \dfrac{0.35}{100} = \dfrac{0.35}{100} \cdot \dfrac{100}{100}$

    $= \dfrac{35}{10,000} = \dfrac{\cancel{5} \cdot 7}{\cancel{5} \cdot 2,000} = \dfrac{7}{2,000}$

43. $0.25\% = \dfrac{0.25}{100} = \dfrac{0.25}{100} \cdot \dfrac{100}{100}$

    $= \dfrac{25}{10,000} = \dfrac{\cancel{25}}{\cancel{25} \cdot 400} = \dfrac{1}{400}$

45. $16\% = 0.16$

47. $81\% = 0.81$

49. $34.12\% = 0.3412$

51. $50.033\% = 0.50333$

53. $6.99\% = 0.0699$

55. $1.3\% = 0.013$

57. $7\frac{1}{4}\% = 7.25\% = 0.0725$

59. $18\frac{1}{2}\% = 18.5\% = 0.185$

61. $460\% = 4.6$

63. $316\% = 3.16$

65. $0.5\% = 0.005$

67. $0.03\% = 0.0003$

69. $0.362 = 36.2\%$

71. $0.98 = 98\%$

73. $1.71 = 171\%$

75. $4 = 400\%$

77. $\frac{2}{5} = 0.4 = 40\%$

79. $\frac{4}{25} = 0.16 = 16\%$

81. $\frac{5}{8} = 0.625 = 62.5\%$

83. $\frac{7}{16} = 0.4375 = 43.75\%$

85. $\frac{9}{4} = 2.25 = 225\%$

87. $\frac{21}{20} = 1.05 = 105\%$

89. $\frac{1}{6} = 0.1666... = 16.7\% = 16\frac{2}{3}\%$

91. $\frac{5}{3} = 1.6666... = 166.7\% = 166\frac{2}{3}\%$

	Fraction	Decimal	Percent
93.	$\frac{157}{5,000}$	0.0314	3.14%
95.	$\frac{51}{125}$	0.408	40.8%
97.	$\frac{21}{400}$	0.0525	$5.25\% = 5\frac{1}{4}\%$
99.	$\frac{7}{3}$	2.33	$233\frac{1}{3}\% \approx 233.3\%$

APPLICATIONS

101. $\frac{91}{100} = 91\%$

103. a. $\frac{6}{50} = \frac{12}{100} = 12\%$

   b. $\frac{12}{50} = \frac{24}{100} = 24\%$

   c. $\frac{2}{50} = \frac{4}{100} = 4\%$

105. a. $7.75\% = 0.075$

   b. $5\% = 0.05$

   c. $14.25\% = 0.1425$

107. Add the rest to get 72.5%. $100 - 72.5 = 27.5\%$

109. a. $\frac{15}{192} = \frac{5 \cdot 3}{64 \cdot 3} = \frac{5}{64}$

   b. 0.078125

   c. 2 right: 7.8125%

111. $33\frac{1}{3}\% = \frac{1}{3} = 0.\overline{3}$

113. a. $\frac{39}{45} = \frac{3 \cdot 13}{3 \cdot 15} = \frac{13}{15}$

b. $13 \div 15 = 86.\overline{6} = 86\frac{2}{3}\% \approx 86.7\%$

115. a. $\frac{1}{4}\%$

b. $\frac{.25}{100} = \frac{25}{10,000} = \frac{1}{400}$

c. $\frac{0.25}{100} = 0.0025\%$

117. $1 \div 365 = .002739... \approx 0.27\%$

WRITING

119. I would choose 25% off, since it seems more significant, being a larger number.

121. It could happen if there are more people there than should be.

REVIEW

123. a. $2(6.5) + 2(10.5) = 13 + 21 = 34 cm$

b. $(6.5)(10.5) = 68.25 cm^2$

Section 6.2: Solving Percent Problems Using Percent Equations and Proportions

VOCABULARY

1. We call "What number is 15% of 25?" a percent sentence. It translates to the percent equation $x = 15\% \cdot 25$.

3. When we find the value of the variable that makes a percent equation true, we say that we have solved the equation.

5. The amount is part of the base. The base is the standard of comparison – it represents the whole of some quantity.

7. The cross products for the proportion $\frac{24}{x} = \frac{36}{100}$ are $24 \cdot 100$ and $x \cdot 36$

CONCEPTS

9. Amount = percent · base

Part = percent · whole

11. $34 + 21 + 16 + 29 = 100\%$

NOTATION

13. a. $12\% = 0.12$

b. $5.6\% = 0.056$

c. $125\% = 1.25$

d. $\frac{1}{4}\% = 0.25\% = 0.0025$

GUIDED PRACTICE

15. $x = 7\% \cdot 16$

a. $\frac{x}{16} = \frac{7}{100}$

$125 = x \cdot 800$

b. $\frac{125}{800} = \frac{x}{100}$

$1 = 94\% \cdot x$

c. $\frac{1}{x} = \frac{94}{100}$

17. $5.4\% \cdot 9 = x$

a. $\frac{x}{9} = \frac{5.4}{100}$

$75.1\% \cdot x = 15$

b. $\frac{15}{x} = \frac{75.1}{100}$

$x \cdot 33.8 = 3.8$

c. $\frac{3.8}{33.8} = \frac{x}{100}$

19. $\dfrac{34}{100} = \dfrac{x}{200}$

$34 \cdot 200 = x \cdot 100$

$\dfrac{6{,}800}{100} = \dfrac{x \cdot 100}{100}$

$68 = x$

21. $\dfrac{88}{100} = \dfrac{x}{150}$

$88 \cdot 150 = x \cdot 100$

$\dfrac{13{,}200}{100} = \dfrac{x \cdot 100}{100}$

$132 = x$

23. $\dfrac{224}{100} = \dfrac{x}{7.9}$

$224 \cdot 7.9 = x \cdot 100$

$\dfrac{1{,}769.6}{100} = \dfrac{x \cdot 100}{100}$

$17.696 = x$

25. $\dfrac{105}{100} = \dfrac{x}{23.2}$

$105 \cdot 23.2 = x \cdot 100$

$\dfrac{2{,}436}{100} = \dfrac{x \cdot 100}{100}$

$24.36 = x$

27. $\dfrac{8}{32} = \dfrac{x}{100}$

$8 \cdot 100 = x \cdot 32$

$\dfrac{800}{32} = \dfrac{x \cdot 32}{32}$

$25\% = x$

29. $\dfrac{51}{60} = \dfrac{x}{100}$

$51 \cdot 100 = x \cdot 60$

$\dfrac{5{,}100}{60} = \dfrac{x \cdot 60}{60}$

$85\% = x$

31. $\dfrac{5}{8} = \dfrac{x}{100}$

$5 \cdot 100 = x \cdot 8$

$\dfrac{500}{8} = \dfrac{x \cdot 8}{8}$

$62.5\% = x$

33. $\dfrac{7}{16} = \dfrac{x}{100}$

$7 \cdot 100 = x \cdot 16$

$\dfrac{700}{16} = \dfrac{x \cdot 16}{16}$

$43.75\% = x$

35. $\dfrac{66}{60} = \dfrac{x}{100}$

$66 \cdot 100 = x \cdot 60$

$\dfrac{6{,}600}{60} = \dfrac{x \cdot 60}{60}$

$110\% = x$

37. $\dfrac{84}{24} = \dfrac{x}{100}$

$84 \cdot 100 = x \cdot 24$

$\dfrac{8{,}400}{24} = \dfrac{x \cdot 24}{24}$

$350\% = x$

39. $\dfrac{9}{x} = \dfrac{30}{100}$

$\dfrac{9}{x} = \dfrac{3}{10}$

$9 \cdot 10 = 3 \cdot x$

$\dfrac{90}{3} = \dfrac{3 \cdot x}{3}$

$30 = x$

41.
$$\frac{36}{x} = \frac{24}{100}$$
$$\frac{36}{x} = \frac{6}{25}$$
$$36 \cdot 25 = 6 \cdot x$$
$$\frac{900}{6} = \frac{6 \cdot x}{6}$$
$$150 = x$$

43.
$$\frac{19.2}{x} = \frac{33\frac{1}{3}}{100}$$
$$19.2 \cdot 100 = \frac{100}{3} \cdot x$$
$$19.2 \cdot \cancel{100} \cdot \frac{3}{\cancel{100}} = \frac{\cancel{3}}{\cancel{100}} \cdot \frac{\cancel{100}}{\cancel{3}} \cdot x$$
$$19.2 \cdot 3 = x$$
$$57.6 = x$$

45.
$$\frac{48.4}{x} = \frac{66\frac{2}{3}}{100}$$
$$48.4 \cdot 100 = \frac{200}{3} \cdot x$$
$$48.4 \cdot \cancel{100} \cdot \frac{3}{\underset{2}{\cancel{200}}} = \frac{\cancel{3}}{\cancel{200}} \cdot \frac{\cancel{200}}{\cancel{3}} \cdot x$$
$$48.4 \cdot \frac{3}{2} = x$$
$$72.6 = x$$

TRY IT YOURSELF

47.
$$\frac{0.5}{40} = \frac{x}{100}$$
$$0.5 \cdot 100 = x \cdot 40$$
$$\frac{50}{40} = \frac{x \cdot 40}{40}$$
$$1.25\% = x$$

49.
$$\frac{7.8}{x} = \frac{12}{100}$$
$$7.8 \cdot 100 = x \cdot 12$$
$$\frac{780}{12} = \frac{x \cdot 12}{12}$$
$$65 = x$$

51.
$$\frac{33}{x} = \frac{33\frac{1}{3}}{100}$$
$$33 \cdot 100 = \frac{100}{3} \cdot x$$
$$33 \cdot \cancel{100} \cdot \frac{3}{\cancel{100}} = \frac{\cancel{3}}{\cancel{100}} \cdot \frac{\cancel{100}}{\cancel{3}} \cdot x$$
$$33 \cdot 3 = x$$
$$99 = x$$

53.
$$\frac{36}{100} = \frac{x}{250}$$
$$36 \cdot 250 = x \cdot 100$$
$$\frac{9{,}000}{100} = \frac{x \cdot 100}{100}$$
$$90 = x$$

55.
$$\frac{16}{20} = \frac{x}{100}$$
$$16 \cdot 100 = x \cdot 20$$
$$\frac{1{,}600}{20} = \frac{x \cdot 20}{20}$$
$$80\% = x$$

57.
$$\frac{0.8}{100} = \frac{x}{12}$$
$$0.8 \cdot 12 = x \cdot 100$$
$$\frac{9.6}{100} = \frac{x \cdot 100}{100}$$
$$0.096 = x$$

59.
$$\frac{3.3}{x} = \frac{7.5}{100}$$
$$3.3 \cdot 100 = 7.5 \cdot x$$
$$\frac{330}{7.5} = \frac{7.5 \cdot x}{7.5}$$
$$44 = x$$

61. $\dfrac{1.25}{0.05} = \dfrac{x}{100}$

$1.25 \cdot 100 = x \cdot 0.05$

$\dfrac{125}{0.05} = \dfrac{x \cdot 0.05}{0.05}$

$2{,}500\% = x$

63. $\dfrac{x}{105} = \dfrac{102}{100}$

$100 \cdot x = 102 \cdot 105$

$\dfrac{100 \cdot x}{100} = \dfrac{10{,}710}{100}$

$x = 107.1$

65. $\dfrac{5.7}{x} = \dfrac{9\frac{1}{2}}{100}$

$5.7 \cdot 100 = \dfrac{19}{2} \cdot x$

$\cancel{570}^{30} \cdot \dfrac{2}{\cancel{19}} = \dfrac{\cancel{2}}{\cancel{19}} \cdot \dfrac{\cancel{19}}{\cancel{2}} \cdot x$

$30 \cdot 2 = x$

$60 = x$

67. $\dfrac{2{,}500}{8{,}000} = \dfrac{x}{100}$

$\dfrac{25}{80} = \dfrac{x}{100}$

$25 \cdot 100 = x \cdot 80$

$\dfrac{2{,}500}{80} = \dfrac{x \cdot 80}{80}$

$31.25\% = x$

69. $\dfrac{7\frac{1}{4}}{100} = \dfrac{x}{600}$

$7.25 \cdot 600 = x \cdot 100$

$\dfrac{43{,}500}{100} = \dfrac{x \cdot 100}{100}$

$43.5 = x$

APPLICATIONS

71. $\dfrac{24}{100} \cdot 50 = 12$: 12K = 12,000 bytes

73. a. $3.80 + 7.50 + 9.45 = \$20.75$

b. $\dfrac{20}{100} \cdot 20.75 = 4.15$: $4.15 rebate

75. $\dfrac{180}{100} \cdot 1.5 = \dfrac{180}{100} \cdot \dfrac{3}{2} = \dfrac{270}{100} = 2.7 in$

77. $\dfrac{28}{40} = \dfrac{x}{100}$

$28 \cdot 100 = 40 \cdot x$

$\dfrac{2{,}800}{40} = \dfrac{40 \cdot x}{40}$

$70 = x$

He passed (barely).

79. $\dfrac{200}{4{,}000} = \dfrac{1}{20} = 0.05 = 5\%$

The driver paid 5% of the cost.

81. $\dfrac{70}{100} = \dfrac{84}{x}$

$70 \cdot x = 84 \cdot 100$

$\dfrac{70x}{70} = \dfrac{8{,}400}{70}$

$x = 120$

83. $\dfrac{33\frac{1}{3}}{100} = \dfrac{4{,}500}{x}$

$\dfrac{100}{3} x = 450{,}000$

$\dfrac{3}{100}\left(\dfrac{100}{3} x\right) = \dfrac{3}{100}(450{,}000)$

$x = 13{,}500 km$

85. $\dfrac{38}{100} \cdot 2{,}700 = 38 \cdot 27 = \$1{,}026 b$

87. $0.25x = 6$

$\dfrac{0.25x}{0.25} = \dfrac{6}{0.25}$

$x = 24$

24 oz. in the larger bottle.

89. Tank 1: $0.5 \cdot 60 = 30 g$

Tank 2: $0.3 \cdot 40 = 12 g$

91. $10,000 = x \cdot 25$

$\dfrac{10,000}{25} = x$

$400 = x$

$40,000\% = x$

93.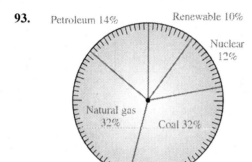

95. $\dfrac{832}{2,600} = 0.32 = 32\%$

$\dfrac{1,118}{2,600} = 0.43 = 43\%$

$\dfrac{338}{2,600} = 0.13 = 13\%$

$\dfrac{156}{2,600} = 0.06 = 6\%$

$\dfrac{156}{2,600} = 0.06 = 6\%$

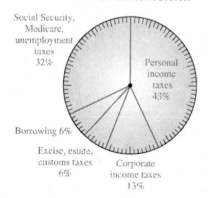

WRITING

97. Out of 20 students in my class, 9 of them are 19 years old. What percent of the students are 19?

99. 150% of a number is larger because 100% represents the number and everything above that is in addition to the original number.

101. It is best to write the number as a fraction when it can be easily converted to a denominator of 100.

REVIEW

103. $2.68 + 6 + 9.09 + 0.3 = 18.17$

105. The distance from 4.9 is 0.1, the distance from 5.001 is 0.001, so 5.001 is closer.

107. $(0.2)^3 = (0.2)(0.2)(0.2) = 0.008$

Section 6.3: Applications of Percent

VOCABULARY

1. Instead of working for a salary or getting paid at an hourly rate, some salespeople are paid on commission. They earn a certain percent of the total dollar amount of the goods or services they sell.

3. a. When we use percent to describe how a quantity has increased compared to its original value we are finding its percent of increase.

   b. When we use percent to describe how a quantity has decreased compared to its original value, we are finding the percent of decrease.

CONCEPTS

5. Sales tax = sales tax rate · purchase price

7. Commission = commission rate · sales

9. a. $59.32 + 4.75 = \$64.07$

   b. $150.99 - 15.99 = \$135.00$

11. To find the percent decrease, subtract the smaller number from the larger number to find the amount of decrease. The find what percent that difference is of the original amount.

GUIDED PRACTICE

13. $0.04 \cdot \$92.70 = \$3.708 \approx \$3.71$

15. $0.05 \cdot \$83.90 = \$4.195 \approx \$4.20$

17. $0.038 \cdot \$68.24 = \$2.59312 \approx \$2.59$
$\$68.24 + \$2.59 = \$70.83$

19. $0.064 \cdot \$60.18 = \$3.85852 \approx \$3.85$
$\$60.18 + \$3.85 = \$64.03$

21. $7.28 = x \cdot 140$
$\dfrac{7.28}{140} = \dfrac{x \cdot 140}{140}$
$0.052 = x$
$x = 5.2\%$

23. $4{,}590 = x \cdot 30{,}000$
$\dfrac{4{,}590}{30{,}000} = \dfrac{x \cdot 30{,}000}{30{,}000}$
$0.153 = x$
$x = 15.3\%$

25. $0.12 \cdot \$95 = \$11.40$

27. $0.35 \cdot \$480 = \$168$

29. $15 = x \cdot 750$
$\dfrac{15}{750} = \dfrac{x \cdot 750}{750}$
$0.02 = x$
$x = 2\%$

31. $20 = x \cdot 500$
$\dfrac{20}{500} = \dfrac{x \cdot 500}{500}$
$0.04 = x$
$x = 4\%$

33. $88 - 80 = 8$
$8 = x \cdot 80$
$\dfrac{8}{80} = \dfrac{x \cdot 80}{80}$
$0.1 = x$
$x = 10\%$

35. $345 - 300 = 45$
$45 = x \cdot 300$
$\dfrac{45}{300} = \dfrac{x \cdot 300}{300}$
$0.15 = x$
$x = 15\%$

37. $30 - 24 = 6$
$6 = x \cdot 30$
$\dfrac{6}{30} = \dfrac{x \cdot 30}{30}$
$0.2 = x$
$x = 20\%$

39. $40 - 36 = 4$
$4 = x \cdot 40$
$\dfrac{4}{40} = \dfrac{x \cdot 40}{40}$
$0.1 = x$
$x = 10\%$

41. $d = 0.33 \cdot \$90 = \$29.70$
$\$90 - \$29.70 = \$60.30$

43. $d = 0.15 \cdot \$58 = \$8.70$
$\$58 - \$8.70 = \$49.30$

45. $d = \$79.95 - \$64.95 = \$15$
$15 = x \cdot 79.95$
$\dfrac{15}{79.95} = \dfrac{x \cdot 79.95}{79.95}$
$0.1876 \approx x$
$x \approx 19\%$

47. $\$209 - \$179 = \$30$
$30 = x \cdot 209$
$\dfrac{30}{209} = \dfrac{x \cdot 209}{209}$
$0.1435 \approx x$
$x \approx 14\%$

APPLICATIONS

49. $0.0595 \cdot \$900 = \$53.55$

51. Subtotal: $\$8.97 + \$9.87 + \$28.50 = \$47.34$

Sales Tax: $0.06 \cdot \$47.34 \approx \$2.84$

Total: $\$47.34 + \$2.84 = \$50.18$

53. $10.32 = x \cdot 129$

$$\frac{10.32}{129} = \frac{x \cdot 129}{129}$$

$0.08 = x$

$x = 8\%$

55. $5 = x \cdot 2{,}000$

$$\frac{5}{2{,}000} = \frac{x \cdot 2{,}000}{2{,}000}$$

$0.0025 = x$

$x = 0.25\%$

57. $0.01 \cdot \$15{,}000 = \$150$

59. Fed. Tax: $28.8 = x \cdot 360$

$$\frac{28.8}{360} = \frac{x \cdot 360}{360}$$

$0.08 = x$

$x = 8\%$

Similarly,

Work. Comp: $\dfrac{13.50}{360} = 3.75\%$

Medicare: $\dfrac{4.32}{360} = 1.2\%$

Social Security: $\dfrac{22.32}{360} = 6.2\%$

61. $84 - 80 = 4$

$4 = x \cdot 80$

$$\frac{4}{80} = \frac{x \cdot 80}{80}$$

$0.05 = x$

$x = 5\%$

63. $7.6 - 5.8 = 1.8$

$1.8 = x \cdot 5.8$

$$\frac{1.8}{5.8} = \frac{x \cdot 5.8}{5.8}$$

$0.3103 \approx x$

$x \approx 31\%$

65. $106 - 42 = 64$

$64 = x \cdot 42$

$$\frac{64}{42} = \frac{x \cdot 42}{42}$$

$1.5238 \approx x$

$x \approx 152\%$

67. $150 - 96 = 54$

$54 = x \cdot 150$

$$\frac{54}{150} = \frac{x \cdot 150}{150}$$

$0.36 = x$

$x = 36\%$

69. $\$112 - \$98 = \$14$

$14 = x \cdot 112$

$$\frac{14}{112} = \frac{x \cdot 112}{112}$$

$0.125 = x$

$x = 12.5\%$

71. $1.25 - 1 = 0.25$

a. $0.25 = 1 \cdot x$

$x = 25\%$

$1.25 - 0.8 = 0.45$

$0.45 = x \cdot 1.25$

b. $\dfrac{0.45}{1.25} = \dfrac{x \cdot 1.25}{1.25}$

$0.36 = x$

$x = 36\%$

73. $0.06 \cdot \$98{,}500 = \$5{,}910$

$\dfrac{1}{2} \cdot \$5{,}910 = \$2{,}955$

75. $37,500 = x \cdot 2,500,000$

$\dfrac{37,500}{2,500,000} = \dfrac{x \cdot 2,500,000}{2,500,000}$

$0.015 = x$

$x = 1.5\%$

77. $144 = x \cdot 160$

$\dfrac{144}{160} = \dfrac{x \cdot 160}{160}$

$0.9 = x$

$x = 90\%$

79. $6 \cdot 6,000 = \$36,000$

$\dfrac{33\frac{1}{3}}{100} \cdot \$36,000 = \dfrac{100}{3} \cdot \dfrac{360}{1} = \$12,000$

81. a. $0.2 \cdot \$39.95 = \$7.99$

b. $\$39.95 - \$7.99 = \$31.96$

83. $5,700 - 5,350 = 350$

$350 = x \cdot 5,700$

$\dfrac{350}{5,700} = \dfrac{x \cdot 5,700}{5,700}$

$0.0614 \approx x$

$x \approx 6\%$

85. $\$399.97 - \$50 = \$349.97$

$50 = x \cdot 399.97$

$\dfrac{50}{399.97} = \dfrac{x \cdot 399.97}{399.97}$

$0.125 \approx x$

$x \approx 13\%$

87. $\$15.48 - \$3.60 = \$11.88$

$3.6 = x \cdot 15.48$

$\dfrac{3.6}{15.48} = \dfrac{x \cdot 15.48}{15.48}$

$0.2326 \approx x$

$x \approx 23\%$

89. $0.55 \cdot \$170 = \$93.50$

$\$170 - \$93.50 = \$76.50$

91. $149.99 = 0.8 \cdot x$

$\dfrac{149.99}{0.8} = \dfrac{0.8 \cdot x}{0.8}$

$\$187.29 \approx x$

WRITING

93. The sales tax is how much money you spend in addition to the list price, while the sales tax rate is what percentage of your purchase you pay in addition to the list price

95. It cannot be used because percent sentences require the base price to be correct.

REVIEW

97. $-5(-5)(-2) = 25(-2) = -50$

99. $-4 - (-7) = -4 + 7 = 3$

101. $|-5 - 8| = |-13| = 13$

Section 6.4: Estimation with Percent

VOCABULARY

1. Estimation can be used to find approximations when exact answers aren't necessary.

CONCEPTS

3. To find 1% of a number, move the decimal point in the number two places to the left.

5. To find 20% of a number, find 10% of the number by moving the decimal point one place to the left, and then double (multiply by 2) the result.

7. To find 25% of a number, divide the number by 4.

9. To find 15% of a number, find the sum of 10% of the number and 5% of the number.

GUIDED PRACTICE

11. a. 2.751

    b. 1% of 300 = 3

13. a. 0.1267

    b. 1% of 10 = 0.1

15. a. 405.9 lb

    b. 10% of 4,000 = 400 lb

17. a. 69.14 min

    b. 10% of 700 = 70 min

19. $0.1 \cdot 350 = 35$

    $2 \cdot 35 = 70$

21. $0.1 \cdot 70 = 7$

    $2 \cdot 7 = 14$

23. $\dfrac{4,200,000}{2} = 2,100,000$

25. $\dfrac{400,000}{2} = 200,000$

27. $\dfrac{16}{4} = 4$

29. $\dfrac{48}{4} = 12$

31. $0.1 \cdot 16,400 = 1,640$

    $\dfrac{1,640}{2} = 820$

33. $0.1 \cdot 400 = 40$

    $\dfrac{40}{2} = 20$

35. $0.1 \cdot 60 = 6$

    $\dfrac{6}{2} = 3$

    $6 + 3 = \$9$

37. $0.1 \cdot 30 = 3$

    $\dfrac{3}{2} = 1.5$

    $3 + 1.5 = \$4.50$

39. $0.1 \cdot 120 = 12$

    $\dfrac{12}{2} = 6$

    $12 + 6 = \$18$

41. $0.1 \cdot 10 = 1$

    $\dfrac{1}{2} = 0.5$

    $1 + 0.5 = \$1.50$

43. $2 \cdot 4 = 8$

45. $2 \cdot 36 = 72$

TRY IT YOURSELF

47. $0.01 \cdot 600 = 6$

    $2 \cdot 6 = 12$

49. $0.1 \cdot 18 = 1.8$

    $3 \cdot 1.8 = 5.4$

51. $3 \cdot 60 = 180$

53. $0.1 \cdot 4,600 = 460$

    $\dfrac{460}{2} = 230$

55. $0.01 \cdot 600 = 6$

57. $0.1 \cdot 120 = 12$

    $\dfrac{12}{2} = 6$

    $12 + 6 = 18$

59. $0.1 \cdot 70 = 7$

61. $\dfrac{280}{4} = 70$

63. $\dfrac{24,000}{2} = 12,000$

65. $2 \cdot 0.9 = 1.8$

67. $0.5 \cdot 98 = 49$

    $0.01 \cdot 49 = 0.49$

69. $0.2 \cdot 400 = 80$

    $0.15 \cdot 80 = 12$

APPLICATIONS

71. $0.1 \cdot 820 = 82$

    $2 \cdot 82 = 164$

    164 students

73. $0.1 \cdot 200 = 20$

    $3 \cdot 20 = 60$

75. $0.1 \cdot 40 = 4$

    $\dfrac{4}{2} = 2$

    $4 + 2 = \$6$

77. $\$28.55 + \$19.75 = \$48.30$

    $0.1 \cdot 50 = 5$

    $\dfrac{5}{2} = 2.5$

    $5 + 2.5 = \$7.50$

79. $\dfrac{120,000}{4} = \$30,000$

81. $2 \cdot 160 = 320\,lb$

83. $0.1 \cdot 700 = 70$

    $3 \cdot 70 = 210$ motorists

85. $24\% \approx 25\%$

    $\dfrac{900}{4} = 225$ people

87. $58\% \approx 60\%$

    $0.1 \cdot 30,000 = 3,000$

    $6 \cdot 3,000 = 18,000$ people

89. $48\% \approx 50\%$

    $\dfrac{6,200}{2} = 3,100$ volunteers

WRITING

91. 200% of a number is twice the number since 100% is the number, 2(100%) is the number 2 times, or twice the number.

93. Since 5% is half of 10%, find 10% of the number and divide by 2.

REVIEW

95. a. $\dfrac{5}{6} + \dfrac{1}{2} = \dfrac{5}{6} + \dfrac{1}{2} \cdot \dfrac{3}{3} = \dfrac{5}{6} + \dfrac{3}{6} = \dfrac{8}{6} = \dfrac{4}{3}$

    b. $\dfrac{5}{6} - \dfrac{1}{2} = \dfrac{5}{6} - \dfrac{1}{2} \cdot \dfrac{3}{3} = \dfrac{5}{6} - \dfrac{3}{6} = \dfrac{2}{6} = \dfrac{1}{3}$

    c. $\dfrac{5}{6} \cdot \dfrac{1}{2} = \dfrac{5 \cdot 1}{6 \cdot 2} = \dfrac{5}{12}$

    d. $\dfrac{5}{6} \div \dfrac{1}{2} = \dfrac{5}{6} \cdot \dfrac{2}{1} = \dfrac{5 \cdot \cancel{2}}{\cancel{2} \cdot 3 \cdot 1} = \dfrac{5}{3}$

Section 6.5: Interest

VOCABULARY

1. In general, <u>interest</u> is money that is paid for the use of money.

3. The percent that is used to calculate the amount of interest to be paid is called the interest <u>rate</u>.

5. The <u>total</u> amount in an investment account is the sum of the principal and the interest.

CONCEPTS

7. a. $125,000

   b. 5%

   c. 30 years

9. a. $7\% = 0.07$

   b. $9.8\% = 0.098$

   c. $6\dfrac{1}{4}\% = 6.25\% = 0.0624$

11. $10,000 \cdot 0.06 \cdot 3 = 600 \cdot 3 = \$1,800$

13. a. compound interest

    b. $1,000

    c. 4

    d. $50

    e. 4 quarters is 1 year

NOTATION

15. $I = Prt$

GUIDED PRACTICE

17. $I = Prt$

$I = \$2,000 \cdot 0.05 \cdot 1$
$= \$100 \cdot 1$
$= \$100$

19. $I = Prt$

$I = \$700 \cdot 0.09 \cdot 4$
$= \$63 \cdot 4$
$= \$252$

21. $I = Prt$

$I = \$500 \cdot 0.025 \cdot 2$
$= \$12.5 \cdot 2$
$= \$25$
$\$500 + \$25 = \$525$

23. $I = Prt$

$I = \$1,500 \cdot 0.012 \cdot 5$
$= \$18 \cdot 5$
$= \$90$
$\$1,500 + \$90 = \$1,590$

25. $\dfrac{9mo}{1} \cdot \dfrac{1yr}{12\,mo} = \dfrac{9}{12}yr = 0.75\,yr$

$I = Prt$

$I = \$550 \cdot 0.04 \cdot 0.75$
$= \$22 \cdot 0.75$
$= \$16.50$

27. $\dfrac{4mo}{1} \cdot \dfrac{1yr}{12\,mo} = \dfrac{9}{12}yr = \dfrac{1}{3}yr$

$I = Prt$

$I = \$1,320 \cdot 0.07 \cdot \dfrac{1}{3}$
$= \$92.4 \cdot \dfrac{1}{3}$
$= \$30.80$

29. $\dfrac{90d}{1} \cdot \dfrac{1yr}{365d} = \dfrac{90}{365}yr = \dfrac{18}{73}yr$

$I = Prt$

$I = \$12,600 \cdot 0.18 \cdot \dfrac{18}{73}$
$= \$2,268 \cdot \dfrac{18}{73}$
$\approx \$559.23$
$\$12,600 + \$559.23 = \$13,159.23$

31. $\dfrac{45d}{1} \cdot \dfrac{1yr}{365d} = \dfrac{45}{365}yr = \dfrac{9}{73}yr$

$I = Prt$

$I = \$40,000 \cdot 0.1 \cdot \dfrac{9}{73}$
$= \$4,000 \cdot \dfrac{9}{73}$
$\approx \$493.15$
$\$40,000 + \$493.15 = \$40,493.15$

33. $A = P\left(1 + \dfrac{r}{n}\right)^{nt}$

$A = 2,000\left(1 + \dfrac{0.03}{4}\right)^{4 \cdot 1}$

$A = 2,000(1.0075)^4$

$A \approx \$2,060.68$

35. $A = P\left(1 + \dfrac{r}{n}\right)^{nt}$

$A = 5,400\left(1 + \dfrac{.04}{4}\right)^{4 \cdot 1}$

$A = 5,400(1.01)^4$

$A \approx \$5619.26$

37.
$$A = P\left(1+\frac{r}{n}\right)^{nt}$$
$$A = 30,000\left(1+\frac{0.048}{365}\right)^{365\cdot 6}$$
$$A = \$40,011.96$$

$\$40,011.96 - \$30,000 = \$10,011.96$ can be withdrawn, leaving the balance intact.

39.
$$A = P\left(1+\frac{r}{n}\right)^{nt}$$
$$A = 55,250\left(1+\frac{0.0855}{365}\right)^{365\cdot 4}$$
$$A \approx \$77,775.64$$

APPLICATIONS

41. $I = \$5,000 \cdot 0.06 \cdot 1 = \$300$
$\$5,000 + \$300 = \$5,300$

43. $I = \$1,200 \cdot 0.055 \cdot 3$
$= \$66 \cdot 3$
$= \$198$

45. $I = \$4,500 \cdot 0.12 \cdot 2$
$= \$540 \cdot 2$
$= \$1,080$
$\$4,500 + \$1,080 = \$5,580$

47. $\dfrac{3mo}{1} \cdot \dfrac{1yr}{12mo} = \dfrac{1}{4}yr = 0.25yr$
$I = \$1,500 \cdot 0.125 \cdot 0.25$
$= \$187.5 \cdot 0.25$
$\approx \$46.88$

49. $\dfrac{30d}{1} \cdot \dfrac{1yr}{365d} = \dfrac{6}{73}yr$
$I = \$4,200 \cdot 0.18 \cdot \dfrac{6}{73}$
$= \$756 \cdot \dfrac{6}{73}$
$\approx \$62.14$
$\$4,200 + \$62.14 = \$4,262.14$

51. $7\dfrac{1}{4}\% = 7.25\% = 0.0725$
$I = \$10,000 \cdot 0.0725 \cdot 2$
$= \$725 \cdot 2$
$= \$1,450$

53. 4. $\$1,200 \cdot 0.08 \cdot 2 = \$192$

5. $\$1,200 + \$192 = \$1,392$

6. $\dfrac{\$1,392}{24} = \$58$

Loan Application Worksheet

1. Amount of loan (principal) $\$1,200.00$
2. Length of loan (time) 2 YEARS
3. Annual percentage rate (simple interest) 8%
4. Interest charged $\$192$
5. Total amount to be repaid $\$1,392$
6. Check method of repayment:
   ☐ 1 lump sum   ☑ monthly payments

Borrower agrees to pay 24 equal payments of $\$58$ to repay loan.

55. $I = \$18m \cdot 0.023 \cdot 3.5 = \$1.449m$
$\$18m + \$1.449m = \$19.449m$

57.
$$A = P\left(1+\frac{r}{n}\right)^{nt}$$
$$A = 600\left(1+\frac{.08}{1}\right)^{3\cdot 1}$$
$$A \approx \$755.83$$

59.
$$A = P\left(1+\frac{r}{n}\right)^{nt}$$
$$A = 1,000\left(1+\frac{0.06}{365}\right)^{365\cdot 4}$$
$$A \approx \$1,271.22$$

61. $A = 545\left(1+\dfrac{0.046}{365}\right)^{365\cdot 1}$

$A \approx \$570.65$

63. $A = 500,000\left(1+\dfrac{0.06}{365}\right)^{365\cdot 1}$

$A \approx \$30,915.66$

65. $A = 90,000\left(1+\dfrac{0.051}{365}\right)^{365\cdot 20}$

$\approx 244,569.75$

The person can withdraw
$\$244,569.75 - \$90,000 = \$154,569.75$ in interest.

WRITING

67. Simple interest is only calculated once. Compound interest is calculated based on previous interest earned / paid, in addition to the principal amount.

69. They would do that because then they do not make interest off of the loan amount.

REVIEW

71. $\sqrt{\dfrac{1}{4}} = \dfrac{1}{2}$

73. $\dfrac{3}{7}+\dfrac{2}{5} = \dfrac{3}{7}\cdot\dfrac{5}{5}+\dfrac{2}{5}\cdot\dfrac{7}{7}$

$= \dfrac{15}{35}+\dfrac{14}{35} = \dfrac{29}{35}$

75. $2\dfrac{1}{2}\cdot 3\dfrac{1}{3} = \dfrac{5}{2}\cdot\dfrac{10}{3} = \dfrac{50}{6}$

$= 8\dfrac{2}{6} = 8\dfrac{1}{3}$

77. $-6^2 = -6\cdot 6 = -36$

Chapter 6 Review

1. $39\%,\ 0.39,\ \dfrac{39}{100}$

3. $100\% - 39\% = 61\%$

5. $\dfrac{15}{100} = \dfrac{3\cdot \cancel{5}}{\cancel{5}\cdot 20} = \dfrac{3}{20}$

7. $\dfrac{9\frac{1}{4}}{100} = \dfrac{9.25}{100} = \dfrac{925}{10,000} = \dfrac{\cancel{25}\cdot 37}{\cancel{25}\cdot 400} = \dfrac{37}{400}$

9. 0.27

11. 6.55

13. 0.0075

15. 83%

17. 5.1%

19. $0.5 = 50\%$

21. $0.875 = 87.5\%$

23. $\dfrac{1}{3} = 0.3333... = 33.3333...\% = 33\dfrac{1}{3}\% \approx 33.3\%$

25. $\dfrac{11}{12} = 0.91666... = 91.666...\% = 91\dfrac{2}{3}\% \approx 91.7\%$

27. a. $97.2\% = 0.972$

b. $\dfrac{97.2}{100} = \dfrac{972}{1,000} = \dfrac{\cancel{4}\cdot 243}{\cancel{4}\cdot 250} = \dfrac{243}{250}$

29. a. $\dfrac{1}{4}\% = 0.25\% = 0.0025$

b. $\dfrac{0.25}{100} = \dfrac{25}{10,000} = \dfrac{\cancel{25}}{\cancel{25}\cdot 400} = \dfrac{1}{400}$

31. a. amount: 15, base: 45, percent: $33\dfrac{1}{3}\%$

b. Amount = percent·base or Part = percent·whole

33. a. $x = 32\%\cdot 96$

b. $64 = x\cdot 135$

c. $9 = 47.2\%\cdot x$

35. $$\frac{x}{500} = \frac{40}{100}$$
$$100 \cdot x = 40 \cdot 500$$
$$\frac{\cancel{100}x}{\cancel{100}} = \frac{20,000}{100}$$
$$x = 200$$

37. $$\frac{1.4}{80} = \frac{x}{100}$$
$$1.4 \cdot 100 = 80 \cdot x$$
$$\frac{140}{80} = \frac{\cancel{80}x}{\cancel{80}}$$
$$1.75 = x$$

1.4 is 1.75% of 80.

39. $A = 2.20 \cdot 55 = 121$

41. $$\frac{7.25}{100} = \frac{43.5}{x}$$
$$7.25 \cdot x = 43.5 \cdot 100$$
$$\frac{7.25x}{7.25} = \frac{4,350}{7.25}$$
$$x = 600$$

43. $0.04 \cdot 15 = 0.6\, gal$

45. $96 = 110 \cdot x$
$$\frac{96}{110} = \frac{\cancel{110}x}{\cancel{110}}$$
$$0.87 \approx x$$

96 is about 87% of 110.

47.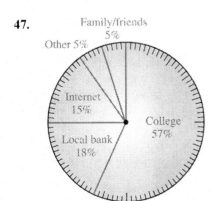

49. $0.055 \cdot 59.99 \approx \$3.30$
$59.99 + 3.30 = \$63.29$

51. Total sales: $\$369.97 + \$299.97 = \$669.94$
$0.06 \cdot \$669.94 \approx \$40.20$

53. $33\frac{1}{3}\% = \frac{1}{3}$
$\frac{1}{3}(12,000)(25) = \$100,000$

55. There was an increase of 3,000 troops.
$$\frac{3,000}{17,000} = 0.1764... \approx 18\%$$

57. a. Sales tax = sales tax rate · purchase price

b. Total cost = purchase price + sales tax

c. Commission = commission rate · sales

59. discount: $\$180$

original price: $\$2,320 + \$180 = \$2,500$

discount rate: $\frac{180}{2,500} = 0.072 = 7.2\%$

61. 3.4023 ; 3

63. 4.34sec ; 4sec

65. 10% is roughly 6, so 20% is roughly 12.

67. 50% of 280,000 is 140,000.

69. 25% of 12 is 3.

71. 10% is roughly 700, so 5% is roughly 350.

73. 200% of 30 is 60.

75. 10% is about $24. 5% is about $12.

$24 + $12 = $36.

77. $50\, fl\, oz \approx 48\, fl\, oz$

25% of 48 is 12 $fl\, oz$

79. Estimate the number of cars with 4,000 and the percent with 5%.

10% of 4,000 is 400 so 5% of 4,000 is 200 cars.

81. $I = P \cdot r \cdot t$

$I = 6{,}000 \cdot 0.08 \cdot 2$

$I = 480 \cdot 2$

$I = \$960$

83. $3\,mo = \dfrac{1}{4}\,yr$

$I = P \cdot r \cdot t$

$I = 2{,}750 \cdot 0.11 \cdot \dfrac{1}{4}$

$I = 302.5 \cdot \dfrac{1}{4}$

$I = \$75.63$

85. a. $I = 1{,}500 \cdot 0.0775 \cdot 1 = \$116.25$

b. $\$1{,}500 + \$116.25 = \$1{,}616.25$

c. $\dfrac{1{,}616.25}{12} = \$134.69$

87. $A = P\left(1 + \dfrac{r}{n}\right)^{nt}$

$A = 5{,}000\left(1 + \dfrac{0.065}{365}\right)^{365 \cdot 3}$

$A \approx \$6{,}076.45$

Chapter 6 Test

1. a. <u>Percent</u> means parts per one hundred.

b. The key words in a percent sentence translate as follows:

* <u>is</u> translates to an equal symbol =

* <u>of</u> translates to multiplication that is shown with a raised dot ·

* <u>what</u> <u>number</u> or <u>what</u> <u>percent</u> translates to an unknown number that is represented by a variable.

c. In the percent sentence "5 is 25% of 20", 5 is the <u>amount</u>, 25% is the <u>percent</u> and 20 is the <u>base</u>.

d. When we use percent to describe how a quantity has increased compared to its original value, we are finding the percent of <u>increase</u>.

e. <u>Simple</u> interest is interest earned only on the original principal. <u>Compound</u> interest is interest paid on the principal and previously earned interest.

3. 199%, $\dfrac{199}{100}$, 1.99

5. a. 0.0006

b. 2.1

c. 0.55375

7. a. 19%

b. 347%

c. 0.5%

9. a. $55\% = \dfrac{55}{100} = \dfrac{\cancel{5} \cdot 11}{\cancel{5} \cdot 20} = \dfrac{11}{20}$

b. $0.01\% = \dfrac{0.01}{100} = \dfrac{1}{10{,}000}$

c. $125\% = \dfrac{125}{100} = \dfrac{5 \cdot \cancel{25}}{4 \cdot \cancel{25}} = \dfrac{5}{4}$

11. a. $\dfrac{1}{30} = \dfrac{1}{30} \cdot \dfrac{100}{1}\% = \dfrac{10}{3}\% = 3\dfrac{1}{3}\% \approx 3.3\%$

b. $\dfrac{16}{9} = \dfrac{16}{9} \cdot \dfrac{100}{1}\% = \dfrac{1{,}600}{9}\% = 177\dfrac{7}{9}\% \approx 177.8\%$

13. $\dfrac{35}{14} = \dfrac{x}{100}$

$35 \cdot 100 = 14 \cdot x$

$\dfrac{3{,}500}{14} = \dfrac{\cancel{14}x}{\cancel{14}}$

$250 = x$

35 is 250% of 14.

15. $\dfrac{18}{x} = \dfrac{20}{100}$

   $18 \cdot 100 = x \cdot 20$

   $\dfrac{1,800}{20} = \dfrac{20x}{20}$

   $90 = x$

17. $A = 2.24 \cdot 60 = 134.4$

19. a. $S = 0.03 \cdot 34 = 1.02\,in$

   b. $34 - 1.02 = 32.98\,in$

21. $2.7 = x \cdot 90$

   $\dfrac{2.7}{90} = \dfrac{90x}{90}$

   $0.03 = x$

   The sales tax rate is 3%.

23. $A = 0.04 \cdot 898 = \$35.92$

25. $I = 0.036 \cdot 40,000 = \$1,440$

   $\$40,000 + \$1,440 = \$41,440$

27. discount: $0.33 \cdot 20 = \$6.60$

   sales price: $\$20 - \$6.6 = \$13.40$

29. a. 20% of 400 is 80.

   b. 50% of 6,000,000 is 3,000,000.

   c. 200% of 20 is 40.

31. Use $30 to estimate the bill.

   10% of 30 is 3, so 5% is 1.5.

   $3 + $1.5 = $4.50

33. $I = P \cdot r \cdot t$

   $I = 3,000 \cdot 0.05 \cdot 1$

   $I = \$150$

35. $I = P \cdot r \cdot t$

   $I = 2,000 \cdot 0.08 \cdot \dfrac{90}{365}$

   $I = 160 \cdot \dfrac{90}{365}$

   $I = \$39.45$

Chapters 1 – 6 Cumulative Review

1. a. six million, fifty-four thousand, three hundred forty-six

   b. 6,000,000 + 50,000 + 4,000 + 300 + 40 + 6

3. $50,055$
   $-7,899$
   $\overline{42,156}$

5. $37\overline{)561}$ = 15 R6

   -37
   191
   -185
   6

7. 1, 2, 4, 5, 8, 10, 20, 40

9. $24 = 2 \cdot 2 \cdot 2 \cdot 3$

   $30 = 2 \cdot 3 \cdot 5$

   $LCM(24, 30) = 2 \cdot 2 \cdot 2 \cdot 3 \cdot 5 = 120$

   $GCF(24, 30) = 2 \cdot 3 = 6$

11. $|-8| \,\square\, -(-5)$

   $8 > 5$

   $|-8| > -(-5)$

13. $75 + 60 + 25 + 3(10) = \$190$

   $\$55 - \$190 = -\$135$

15. a. undefined

    b. 0

    c. 0

    d. $0-(-14)=0+14=14$

17. $-5,700+(-2,300)+3,400+2,700$
    $=-8,000+6,100$
    $=-1,900$

19. $\dfrac{4}{5}=\dfrac{4}{5}\cdot\dfrac{9}{9}=\dfrac{36}{45}$

21. $A=\dfrac{1}{2}bh=\dfrac{1}{2}\cdot 50\cdot 26$
    $=25\cdot 26=650\,in^2$

23. $\dfrac{9}{10}-\dfrac{3}{14}=\dfrac{9}{10}\cdot\dfrac{7}{7}-\dfrac{3}{14}\cdot\dfrac{5}{5}$
    $=\dfrac{63}{70}-\dfrac{15}{70}=\dfrac{48}{70}$
    $=\dfrac{\cancel{2}\cdot 24}{\cancel{2}\cdot 35}=\dfrac{24}{35}$

25. $\dfrac{1}{3}-\dfrac{1}{4}=\dfrac{1}{3}\cdot\dfrac{4}{4}-\dfrac{1}{4}\cdot\dfrac{3}{3}=\dfrac{4}{12}-\dfrac{3}{12}=\dfrac{1}{12}\,lb$

27. $32+\dfrac{3}{4}\cdot 5=32+\dfrac{15}{4}=32+3\dfrac{3}{4}=35\dfrac{3}{4}\,in$

29. a. 452.03

    b. 452.030

31. $40(15.90)+4(1.5)(15.90)$
    $=\$636+\$95.4=\$731.40$

33. $\begin{array}{r}0.733\\15\overline{)11.000}\\-\underline{105}\\50\\-\underline{45}\\50\\-\underline{45}\\5\end{array}$

    $\dfrac{11}{15}=0.7\overline{3}$

35. $\dfrac{1\tfrac{1}{4}}{1\tfrac{1}{2}}=\dfrac{\tfrac{5}{4}}{\tfrac{3}{2}}=\dfrac{5}{4}\cdot\dfrac{2}{3}=\dfrac{5\cdot\cancel{2}}{2\cdot\cancel{2}\cdot 3}=\dfrac{5}{6}$

37. $\dfrac{960hr}{1}\cdot\dfrac{1\,day}{24hr}=\dfrac{960}{24}=40\,days$

39. $\dfrac{6.5kg}{1}\cdot\dfrac{2.2lb}{1kg}=6.5\cdot 2.2lb=14.3lb$

41. $20=0.16\cdot x$
    $\dfrac{20}{0.16}=\dfrac{\cancel{0.16}x}{\cancel{0.16}}$
    $125=x$

43. There was an increase of 234.
    $\dfrac{234}{300}=0.78=78\%$

45. Estimate $80 for the bill.

    10% of 80 is 8 so 5% of 80 is 4.

    $8 + $4 = $12

    Total check: $78.18 + $12 = $90.18

CHAPTER 7: Graphs and Statistics

Section 7.1: Reading Graphs and Tables

VOCABULARY

1. (a)

3. (c)

5. (d)

7. A horizontal or vertical line used for reference in a graph is called an <u>axis</u>.

CONCEPTS

9. To read a table, we must find the <u>intersection</u> of the row and column that contains the desired information.

11. A pictograph is like a bar graph, but the bars are made from <u>pictures</u> or symbols.

13. A histogram is a bar graph with three important features:

    * The <u>bars</u> of a histogram touch.

    * Data values never fall at the <u>edge</u> of a bar.

    * The widths of the bars of a histogram are <u>equal</u> and represent a range of values.

NOTATION

15. Half, or about 500 buses.

GUIDED PRACTICE

17. Row 8: $10.70

19. 2 pounds 9 ounces alone: $9.90

    3 pounds 8 ounces alone: $11.95

    Shipped separately: $9.90 + $11.95 = $21.85

    5 pounds 17 ounces = 6 pounds 1 ounce together: $17.30

    She would save $21.85 - $17.30 = $4.55

21. fish, cat, dog

23. No: There are roughly 175 million cats and dogs, more than 175 million fish.

25. yes

27. about 3,000,000 + 3,000,000 + 4,000,000 = 10,000,000 metric tons

29. 1990, 2000, 2007

31. 11,000,000 − 7,000,000 = 4,000,000 metric tons

33. seniors

35. $7.5 \cdot 100 - 7 \cdot 100 = 750 - 700 = \$50$

37. Chinese

39. Combined: 5 + 1 + 1 + 2 + 5 = 14% - no.

41. 5 + 2 + 18 + 3 + 3 + 5 + 1 + 1 = 38%

    100 − 38 = 62%

43. $$\frac{18}{100} = \frac{x}{6,771,000,000}$$
    $$18 \cdot 6,771,000,000 = x \cdot 100$$
    $$\frac{121,878,000,000}{100} = \frac{x \cdot 100}{100}$$
    $$1,219,000,000 \approx x$$

45. 493

47. 2002 to 2003

    2004 to 2005

    2005 to 2006

    2007 to 2008

49. 2001 and 2003

51. between 2005 and 2006: a decrease of 492 − 478 = 14 resorts

53. runner number 1

55. B

57. runner number 1 (distance from the start decreases)

59. Runner 1 was running, runner 2 was catching a break.

61. a. 27

b. $13 + 9 = 22$

TRY IT YOURSELF

63.
$16,056.25 + 0.28(79,250 - 78,850)$
$= 16,056.25 + 0.28(400)$
$= 16,056.25 + 112$
$= \$16,168.25$

65. a.
$4,481.25 + 0.25(53,000 - 32,550)$
$= 4,481.25 + 0.25(20,450)$
$= 4,481.25 + 5,112.50$
$= \$9593.75$

b.
$1,605 + 0.15(51,000 - 16,050)$
$= 1,605 + 0.15(34,950)$
$= 1,605 + 5,242.50$
$= \$6,847.50$

c. $9593.75 - 6847.50 = \$2,746.25$

67. 2000; about 3.3%

69. It increased by about 1%.

71. The altitude increased.

73. D

75. reckless driving and failure to yield

77. reckless driving

79. about $440

81. no

83. the miner's (graph is steeper)

85. Miners: $930 - 440 = \$490$

Construction: $800 - 390 = \$410$

The miners received the greater increase.

87. $4.2 \cdot 10 = \$42$

89. $3.3 \cdot 10 \cdot 5 - 2.8 \cdot 10 \cdot 5 = 165 - 140 = \$25$

91. $\dfrac{8}{71} \approx 0.11 = 11\%$

93. $\dfrac{8+7}{71} = \dfrac{15}{71} \approx 0.21 = 21\%$

95.

97.

WRITING

99. a. I would use a circle graph as it makes it easier to make multiple comparisons.

b. I would use a line graph so I could judge the trend over time.

c. I would use a histogram to compare the time spent for various exams.

d. I would use a table since there are several different categories.

e. I would use a histogram so I could compare the salaries better.

REVIEW

101. 11, 13, 17, 19, 23, 29

103. 0 and 4

Section 7.2: Mean, Median, and Mode

VOCABULARY

1. The <u>mean</u> (average) of a set of values is the sum of the values divided by the number of values in the set.

3. The <u>mode</u> of a set of values is the single value that occurs most often.

CONCEPTS

5. $\text{Mean} = \dfrac{\text{the sum of the values}}{\text{the number of values}}$

7. a. an even number

   b. 6 and 8

   c. $\text{Median} = \dfrac{6+8}{2} = \dfrac{14}{2} = 7$

GUIDED PRACTICE

9. $\dfrac{3+4+7+7+8+11+16}{7}$
$= \dfrac{56}{7} = 8$

11. $\dfrac{5+9+12+35+37+45+60+77}{8}$
$= \dfrac{280}{8} = 35$

13. $\dfrac{15+17+12+19+27+17+19+35+20}{9}$
$= \dfrac{171}{9} = 19$

15. $\dfrac{4.2+3.6+7.1+5.9+8.2}{5}$
$= \dfrac{29}{5} = 5.8$

17. 1 2 5 **9** 11 17 29

    Median = 9

19. 1 3 4 4 **5** 6 7 7 7

    Median = 5

21. 13.6 15.1 **17.2** 19.7 44.9

    Median = 17.2

23. $\dfrac{1}{100} \quad \dfrac{1}{3} \quad \dfrac{5}{8} \quad \dfrac{999}{1{,}000} \quad \dfrac{16}{15}$

    Median = $\dfrac{5}{8}$

25. 6 7 **8 10** 16 63

    Median = $\dfrac{8+10}{2} = \dfrac{18}{2} = 9$

27. 1 39 **41 47** 50 51

    Median = $\dfrac{41+47}{2} = \dfrac{88}{2} = 44$

29. 1.7 1.8 2.0 **2.0 2.1** 2.1 2.3 9.0

    Median = $\dfrac{2.0+2.1}{2} = \dfrac{4.1}{2} = 2.05$

31. $\dfrac{1}{5} \quad \dfrac{2}{5} \quad \dfrac{3}{5} \quad \dfrac{7}{5} \quad \dfrac{11}{5} \quad \dfrac{13}{5}$

    Median = $\dfrac{\frac{3}{5}+\frac{7}{5}}{2} = \dfrac{\frac{10}{5}}{2} = \dfrac{2}{2} = 1$

33. 3

35. -6

37. 22.7

39. both $\dfrac{1}{3}$ and $\dfrac{1}{2}$ (bimodal)

APPLICATIONS

41. a. $\dfrac{75+80+90+85}{4}$
$= \dfrac{330}{4} = 82.5$

b. $\dfrac{75+80+90+2\cdot 85}{5}$
$=\dfrac{415}{5}=83$

43. a. $\dfrac{98,790}{37}=2,670\,mi$

b. $\dfrac{2,670\,mi}{30d}=89\,mi/d$

45. a. $2,500+4\cdot 500+35\cdot 150+85\cdot 25$
$=2,500+2,000+5,250+2,125$
$=\$11,875$

b. $1+4+35+85=125$

c. $\dfrac{11,875}{125}=\$95$

47. a. $\dfrac{sum}{20}=\dfrac{1300}{20}=65¢$

b. $45, 50, 50, 50, 50, 50, 50, 50, 60, 60,$
$60, 70, 75, 75, 75, 75, 75, 80, 100, 100$

Median = $60¢$

c. Mode = $50¢$

49. $\dfrac{53+54+\cdots 68+71}{12}$
$=\dfrac{732}{12}=61°$

51. $\dfrac{5\cdot 2+3\cdot 4+1\cdot 2+4\cdot 1}{13}$
$=\dfrac{10+12+2+4}{13}$
$=\dfrac{29}{13}\approx 2.23$

53. $\dfrac{3\cdot 1+4\cdot 2+1\cdot 4+3\cdot 2+5\cdot 2}{16}$
$=\dfrac{3+8+4+6+10}{16}$
$=\dfrac{31}{16}\approx 2.5$

55. 85 85 85 85 85

Mean, Median, and Mode = 85

57. $B: \dfrac{37+53+78}{3}=\dfrac{168}{3}=56$

$S: \dfrac{53+57+58}{3}=\dfrac{168}{3}=56$

They had the same average, but the sister was more consistent.

59. 11.2 17.5 18 **24 26** 27 28 28.5

Median = $\dfrac{24+26}{2}=\dfrac{50}{2}=25\,oz$

Mean = $\dfrac{11.2+17.5+\cdots 28.5}{8}$
$=\dfrac{180.2}{8}=22.525\,oz$

61. Mean = $\dfrac{6.6+6.4+\cdots 6.3}{15}$
$=\dfrac{102.6}{15}\approx 6.8$

6.0 6.3 6.3 6.4 6.4 6.6 6.6

6.9 7.0 7.0 7.2 7.3 7.3 7.4 7.9

Median = 6.9

63. 3 3 4 4 4 4 5 **5 5** 6 6 6 7 8 9 9 9 13

Median = 5lb

Mode = 4lb

WRITING

65. The mean is the sum of the values divided by the number of the values. The mode is the value which appears the most often. The median is the middle value (or the average of the two middle values) when the values are written in numerical order.

67. Median would be the best way to evaluate it since the ginormous CEO salaries do not skew the results high.

REVIEW

69.
$$\frac{x}{100} = \frac{52}{80}$$
$$x \cdot 80 = 52 \cdot 100$$
$$\frac{x \cdot 80}{80} = \frac{5,200}{80}$$
$$x = 65\%$$

71.
$$\frac{28}{x} = \frac{66\frac{2}{3}}{100}$$
$$28 \cdot 100 = \frac{200}{3} \cdot x$$
$$28 \cdot 100 \cdot \frac{3}{200} = \frac{3}{200} \cdot \frac{200}{3} \cdot x$$
$$28 \cdot \frac{3}{2} = x$$
$$42 = x$$

73.
$$\frac{5}{8} = \frac{x}{100}$$
$$5 \cdot 100 = x \cdot 8$$
$$\frac{500}{8} = \frac{x \cdot 8}{8}$$
$$62.5\% = x$$

75.
$$\frac{7\frac{1}{4}}{100} = \frac{x}{600}$$
$$7.25 \cdot 600 = x \cdot 100$$
$$\frac{43,500}{100} = \frac{x \cdot 100}{100}$$
$$43.5 = x$$

Chapter 7 Review

1. a. -18°

b. -71°

3. 20

5. Germany and India with about 17.

7. 29%

9. women

11. about 4,100 animals

13. roughly 3 elephants worth, or 3,000 animals

15. oxygen

17. $0.10(135) = 13.5 lb$

19. about 3,000 million eggs

21. 2007; about 2,975 million eggs

23. between 2006 and 2007

25. about $3,075 - 2,775 = 300$ million eggs

27. 60

29. $80 + 50 + 30 = 160$

31. $\frac{87+92+97+100+100+98+90+98}{8}$
$= \frac{762}{8} = 95.25$

He earned an A in the class.

33. $1.2 oz$ happened most often.

35. mean: $\frac{7.8 + \cdots 7.5}{9} = \frac{65.7}{9} = 7.3$

median:
$6.7, 6.8, 6.9, 6.9, 7.2, 7.5, 7.8, 7.9, 8.0$: 7.2

mode: 6.9

37. $\dfrac{5\cdot 20+10\cdot 65+20\cdot 25+50\cdot 5+100\cdot 10}{20+65+25+5+10}$

$=\dfrac{2{,}500}{125}=\$20$

Chapter 7 Test

1. a. A horizontal or vertical line used for reference in a bar graph is called an <u>axis</u>.

b. The <u>mean</u> (average) of a set of values is the sum of the values divided by the number of values in the set.

c. The <u>median</u> of a set of values written in increasing order is the middle value.

d. The <u>mode</u> of a set of values is the single value that occurs most often.

e. The mean, median, and mode are three measures of <u>central</u> tendency.

3. a. love seat; 150 feet

b. 140 – 90 = 50 feet more

c. 60 + 100 + 90 + 90 = 340 feet

5. a. about 38 grams

b. about 40 – 25 = 15 grams

7. a. about 27,000 officers

b. 1989; about 26,000 officers

c. 2000; about 41,000 officers

d. about 41,000 – 36,000 = 5,000 officers

9. a. 22 employees

b. 8 + 22 = 30 employees

c. 28 + 20 + 9 = 57 employees

11. $\dfrac{5\cdot 3+4\cdot 4+3\cdot 5+2\cdot 6+1\cdot 2}{3+4+5+6+2}$

$=\dfrac{60}{20}=3\,\text{stars}$

13. mean: $\dfrac{5.39+\cdots+3.93}{7}=\dfrac{30.87}{7}=4.41m$

median: the middle value is $4.25m$

mode: the most common value is $4.25m$

Chapters 1 – 7 Cumulative Review

1. Fifty-two million, nine hundred forty thousand, five hundred fifty-nine

50,000,000 + 2,000,000 + 900,000 + 40,000 + 500 + 50 + 9

3.
```
  1 1  1 1
  38,908
 +15,696
  ──────
  54,604
```

5.
```
     345
    × 67
    ────
   2,415
 +20,700
   ─────
  23,115
```

7. Verify by addition that $683+459=1{,}142$.

9. 1999 + 12 = 2011

11. 2, 3, 5, 7, 11, 13, 17, 19, 23, 29

13. $15+5\left[12-\left(2^{2}+4\right)\right]$

$=15+5\left[12-(4+4)\right]$

$=15+5[12-8]$

$=15+5[4]$

$=15+20$

$=35$

15.

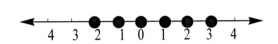

17. a. $-25+5=-20$

b. $25-(-5)=25+5=30$

c. $-25(5)=-125$

d. $\dfrac{-25}{-5}=\dfrac{25}{5}=5$

19. $\dfrac{(-6)^2-1^5}{-4-3}=\dfrac{36-1}{-7}=\dfrac{35}{-7}=-5$

21. $-\left|\dfrac{45}{-9}-(-9)\right|=-|-5+9|=-|4|=-4$

23. a. $\dfrac{60}{108}=\dfrac{5\cdot\cancel{12}}{9\cdot\cancel{12}}=\dfrac{5}{9}$

b. $\dfrac{24}{16}=\dfrac{3\cdot\cancel{8}}{2\cdot\cancel{8}}=\dfrac{3}{2}$

25. $\dfrac{4}{5}\cdot\dfrac{2}{7}=\dfrac{8}{35}$

27. $\dfrac{1}{2}-\dfrac{2}{3}=\dfrac{1}{2}\cdot\dfrac{3}{3}-\dfrac{2}{3}\cdot\dfrac{2}{2}=\dfrac{3}{6}-\dfrac{4}{6}=-\dfrac{1}{6}$

29. $\dfrac{7}{15}\cdot 300=\dfrac{7\cdot\cancel{15}\cdot 20}{\cancel{15}}=140$ minutes in the lab per week.

$300-140=160$ minutes in lecture per week.

31. $26-19\dfrac{1}{4}=25\dfrac{4}{4}-19\dfrac{1}{4}=6\dfrac{3}{4}$ in

33. $\dfrac{-\dfrac{1}{5}}{\dfrac{8}{15}}=-\dfrac{1}{5}\cdot\dfrac{15}{8}=-\dfrac{3\cdot\cancel{5}}{\cancel{5}\cdot 8}=-\dfrac{3}{8}$

35.
$$\begin{array}{r}
\overset{1\ 2\ 2}{213.16}\\
1,504.80\\
89.73\\
+\ \ \ \ 7.50\\
\hline
\$1,815.19
\end{array}$$

37.
$$\begin{array}{r}
7\cancel{6}\cancel{0}.\cancel{2}\\
-6\ 1\ 4\ .\ 7\\
\hline
1\ 4\ 5\ .\ 5
\end{array}$$

39. Begin by moving the decimal place 2 units right.

$$\begin{array}{r}
745\\
72\overline{)53640}\\
-504\ \ \\
\hline
324\ \ \\
-288\ \ \\
\hline
360\\
-360\\
\hline
0
\end{array}$$

41.
$$\begin{array}{r}
0.7272\\
11\overline{)8.0000}\\
-77\ \ \ \ \ \\
\hline
30\ \ \ \ \\
-22\ \ \ \ \\
\hline
80\ \ \\
-77\ \ \\
\hline
30\\
-22\\
\hline
8
\end{array}$$

$\dfrac{8}{11}=0.\overline{72}$

43. $\dfrac{8\,ft}{4\,yd} = \dfrac{8\,ft}{12\,ft} = \dfrac{2 \cdot \cancel{4\,ft}}{3 \cdot \cancel{4\,ft}} = \dfrac{2}{3}$

45. $\dfrac{\tfrac{7}{8}}{\tfrac{1}{2}} = \dfrac{\tfrac{1}{4}}{x}$

$\dfrac{7}{8} \cdot x = \dfrac{1}{2} \cdot \dfrac{1}{4}$

$\dfrac{7x}{8} = \dfrac{1}{8}$

$7x = 1$

$\dfrac{7x}{7} = \dfrac{1}{7}$

$x = \dfrac{1}{7}$

47. $\dfrac{640\,cm}{1} \cdot \dfrac{1\,m}{100\,cm} = \dfrac{640}{100}m = 6.4\,m$

49.

Fraction	Decimal	Percent
$\dfrac{3}{100}$	0.03	3%
$\dfrac{225}{100} = \dfrac{9 \cdot 25}{4 \cdot 25} = \dfrac{9}{4}$	2.25	225%
$\dfrac{41}{1,000}$	0.041	4.1%

51. $x = 1.05(23.2) = 24.36$

53. $0.08 \cdot 98.95 \approx \$7.92$

55. Estimate $80 for the bill.

10% of 80 is 8 so 5% of 80 is 4.

$8 + $4 = $12

57. $I = P \cdot r \cdot t$

$I = 12,600 \cdot 0.18 \cdot \dfrac{90}{365}$

$I = 2,268 \cdot \dfrac{90}{365}$

$I \approx \$559.23$

$\$12,600 + \$559.23 = \$13,159.23$

59. a. 36 in 2008

b. an increase of 16 deaths from 2007 to 2008

c. a decrease of 8 deaths from 2008 to 2009

CHAPTER 8: An Introduction to Algebra

8.1: The Language of Algebra

VOCABULARY

1. <u>Variables</u> are letters (or symbols) that stand for numbers.

3. Variables and/or numbers can be combined with the operations of arithmetic to create algebraic <u>expressions</u>.

5. Addition symbols separate algebraic expressions into parts called <u>terms</u>.

7. The <u>coefficient</u> of the term $10x$ is 10.

CONCEPTS

9. $(12-h)\,in$

11. a. $(x+20)\,oz$

    b. $(100-p)\,lb$

NOTATION

13. $9a - a^2 = 9(5) - 5^2$
    $= 9(5) - 25$
    $= 45 - 25$
    $= 20$

15. $4x$

17. $2w$

GUIDED PRACTICE

19. a. $x+y=y+x$

    b. $(r+s)+t=r+(s+t)$

21. $0 \cdot s = 0 \quad s \cdot 0 = 0$

23. a. 4

    b. 3, 11, -1, 9

25.

$6m$	$-75t$	w	$\frac{1}{2}bh$	$\frac{x}{5}$	t
6	-75	1	$\frac{1}{2}$	$\frac{1}{5}$	1

27. term

29. factor

31. $l+15$

33. $50x$

35. $\dfrac{w}{l}$

37. $P + \dfrac{2}{3}P$

39. $k^2 - 2{,}005$

41. $2a - 1$

43. $\dfrac{1{,}000}{n}$

45. $2p + 90$

47. $3(35 + h + 300)$

49. $p - 680$

51. $4d - 15$

53. $2(200 + t)$

55. $|a - 2|$

57. $\dfrac{1}{10}d$

59. three-fourths of r

61. 50 less than t

63. the product of x, y, and z

65. twice m, increased by 5

67. $(x+2)$ in

69. $(36-x)$ in

71. $-y = -(-2) = 2$

73. $-z + 3x = -(-4) + 3(3) = 4 + 9 = 13$

75. $3y^2 - 6y - 4 = 3(-2)^2 - 6(-2) - 4$
    $= 3 \cdot 4 + 12 - 4 = 12 + 12 - 4$
    $= 24 - 4 = 20$

77. $(3+x)y = (3+3)(-2) = 6(-2) = -12$

79. $(x+y)^2 - |z+y| = (3+(-2))^2 - |-4+(-2)|$
    $= 1^2 - |-6| = 1 - 6 = -5$

81. $-\dfrac{2x+y^3}{y+2z} = -\dfrac{2(3)+(-2)^3}{-2+2(-4)}$
    $= -\dfrac{6+(-8)}{-2-8} = -\dfrac{-2}{-10} = -\dfrac{1}{5}$

83. $b^2 - 4ac = 5^2 - 4(-1)(-2)$
    $= 25 + 4(-2) = 25 - 8 = 17$

85. $a^2 + 2ab + b^2 = (-5)^2 + 2(-5)(-1) + (-1)^2$
    $= 25 + 10 + 1 = 36$

87. $\dfrac{n}{2}[2a + (n-1)d] = \dfrac{10}{2}[2(-4.2) + (10-1)6.6]$
    $= 5[-8.4 + 9(6.6)] = 5[-8.4 + 59.4]$
    $= 5[51] = 255$

89. $(27c^2 - 4d^2)^3 = \left(27\left(\dfrac{1}{3}\right)^2 - 4\left(\dfrac{1}{2}\right)^2\right)^3$
    $= \left(27 \cdot \dfrac{1}{9} - 4 \cdot \dfrac{1}{4}\right)^3 = (3-1)^3 = 2^3 = 8$

APPLICATIONS

91. a. Let x = the weight of the Element (in pounds). Then $2x - 340$ = the weight of the Hummer (in pounds).

    b. $2 \cdot 3{,}370 - 340 = 6{,}400 lb$

93. a. Let x = the age of Apple. Then $x + 80$ = the age of IBM and $x - 9$ = the age of Dell.

    b. IBM: $32 + 80 = 112 yr$
       Dell: $32 - 9 = 23 yr$

WRITING

95. A variable is a letter used to represent a number. One use is for when the actual value of the number is unknown.

97. $2 < x$ would translate into the statement "2 is less than x" while "2 less than x" would translate into $x - 2$.

REVIEW

99. $12 = 2 \cdot 2 \cdot 3$
    $15 = 3 \cdot 5$
    $LCD(12, 15) = 2 \cdot 2 \cdot 3 \cdot 5 = 60$

101. $\left(\dfrac{2}{3}\right)^3 = \left(\dfrac{2}{3}\right)\left(\dfrac{2}{3}\right)\left(\dfrac{2}{3}\right)$
     $= \dfrac{2 \cdot 2 \cdot 2}{3 \cdot 3 \cdot 3} = \dfrac{8}{27}$

8.2: Simplifying Algebraic Expressions

VOCABULARY

1. To simplify the expression $5(6x)$ means to write it in a simpler form: $5(6x) = 30x$.

3. To perform the multiplication $2(x+8)$, we use the distributive property.

5. Terms such as $7x^2$ and $5x^2$, which have the same variables raised to exactly the same power, are called like terms.

CONCEPTS

7. a. $4(9t) = (4 \cdot 9)t = 36t$

b. associative property of multiplication

9. a. $2(x+4) = 2x+8$

b. $2(x-4) = 2x-8$

c. $-2(x+4) = -2x-8$

d. $-2(x-4) = -2x+8$

11. a. $5(2x) = 10x$

b. can't be simplified

c. $6(-7x) = -42x$

d. can't be simplified

e. $2(3x)(3) = 6x(3) = 18x$

f. $2+3x+3 = 3x+5$

NOTATION

13. a. $6(h-4)$

b. $-(z+16)$

GUIDED PRACTICE

15. $3 \cdot 4t = (3 \cdot 4)t = 12t$

17. $9(7m) = (9 \cdot 7)m = 63m$

19. $5(-7q) = (5(-7))q = -35q$

21. $5t \cdot 60 = (5 \cdot 60)t = 300t$

23. $(-5.6x)(-2) = (-5.6(-2))x = 11.2x$

25. $5(4c)(3) = (5 \cdot 4 \cdot 3)c = 60c$

27. $-4(-6)(-4m) = (-4(-6)(-4))m = -96m$

29. $\dfrac{5}{3} \cdot \dfrac{3}{5}g = \dfrac{15}{15}g = 1g = g$

31. $12\left(\dfrac{5}{12}x\right) = \cancel{12}\left(\dfrac{5}{\cancel{12}}x\right) = 5x$

33. $8\left(\dfrac{3}{4}y\right) = \overset{2}{\cancel{8}}\left(\dfrac{3}{\cancel{4}}y\right) = 2(3y) = 6$

35. $5(x+3) = 5x + 5 \cdot 3 = 5x + 15$

37. $-3(4x+9) = -3(4x)+(-3)(9)$
$= -12x - 27$

39. $45\left(\dfrac{x}{5}+\dfrac{2}{9}\right) = \overset{9}{\cancel{45}} \cdot \dfrac{x}{\cancel{5}} + \overset{5}{\cancel{45}} \cdot \dfrac{2}{\cancel{9}}$
$= 9x + 5 \cdot 2 = 9x + 10$

41. $0.4(x+4) = 0.4x + 0.4(4) = 0.4x + 1.6$

43. $6(6c-7) = 6 \cdot 6c - 6 \cdot 7 = 36c - 42$

45. $-6(13c-3) = -6 \cdot 13c - (-6)(3)$
$= -78c + 18$

47. $-15(-2t-6) = (-15)(-2t)-(-15)(6)$
$= 30t + 90$

49. $-1(-4a+1) = (-1)(-4a)+(-1)1$
$= 4a - 1$

51. $(3t+2)8 = 3t(8)+2(8) = 24t+16$

53. $(3w-6)\dfrac{2}{3} = \cancel{3}w\left(\dfrac{2}{\cancel{3}}\right) - \overset{2}{\cancel{6}}\left(\dfrac{2}{\cancel{3}}\right)$
$= 2w - 3 \cdot 2 = 2w - 6$

55. $4(7y+4)2 = 8(7y+4) = 8 \cdot 7y + 8 \cdot 4$
$= 56y + 32$

57. $25(2a-3b+1) = 25 \cdot 2a - 25 \cdot 3b + 25 \cdot 1$
$= 50a - 75b + 25$

59. $-(x-7) = -1(x-7) = -1 \cdot x - (-1)7$
    $= -x + 7$

61. $-(-5.6y + 7) = -1(-5.6y + 7)$
    $= -1(-5.6y) + (-1)7$
    $= 5.6y - 7$

63. $3x, 2x$

65. $-3m^3, -m^3$

67. $3x + 7x = 10x$

69. $-4x + 4x = 0x = 0$

71. $-7b^2 + 27b^2 = 20b^2$

73. $13r - 12r = 1r = r$

75. $36y + y - 9y = 37y - 9y = 28y$

77. $43s^3 - 44s^3 = -1s^3 = -s^3$

79. $-9.8c + 6.2c = -3.6c$

81. $-0.2r - (-0.6r) = -0.2r + 0.6r = 0.4r$

83. $\dfrac{3}{5}t + \dfrac{1}{5}t = \dfrac{4}{5}t$

85. $-\dfrac{7}{16}t - \dfrac{3}{16}t = -\dfrac{10}{16}t = -\dfrac{5}{8}t$

87. $15y - 10 - y - 20y = 14y - 10 - 20y$
    $= -6y - 10$

89. $3x + 4 - 5x + 1 = -2x + 5$

91. does not simplify (no like terms)

93. $4x^2 + 5x - 8x + 9 = 4x^2 - 3x + 9$

95. $2z + 5(z - 3) = 2z + 5z - 15 = 7z - 15$

97. $2(s^2 - 7) - (s^2 - 2) = 2s^2 - 14 - s^2 + 2$
    $= s^2 - 12$

99. $-9(3r - 9) - 7(2r - 7) = -27r + 81 - 14r + 49$
    $= -41r + 130$

101. $36\left(\dfrac{2}{9}x - \dfrac{3}{4}\right) + 36\dfrac{1}{2} = \cancel{36}^{4} \cdot \dfrac{2}{\cancel{9}}x - \cancel{36}^{9} \cdot \dfrac{3}{\cancel{4}} + \cancel{36}^{18}$
     $= 4 \cdot 2x - 9 \cdot 3 + 18 \cdot 1 = 8x - 27 + 18 = 8x - 9$

TRY IT YOURSELF

103. $6 - 4(-3c - 7) = 6 + 12c + 28 = 12c + 34$

105. $-4r - 7r + 2r - r = -11r + 2r - r$
     $= -9r - r = -10r$

107. $24\left(-\dfrac{5}{6}r\right) = -\cancel{24}^{4} \cdot \dfrac{5}{\cancel{6}}r$
     $= -4 \cdot 5r = -20r$

109. $a + a + a = 3a$

111. $60\left(\dfrac{3}{20}r - \dfrac{4}{15}\right) = \cancel{60}^{3} \cdot \dfrac{3}{\cancel{20}}r - \cancel{60}^{4} \cdot \dfrac{4}{\cancel{15}}$
     $= 3 \cdot 3r - 4 \cdot 4 = 9r - 16$

113. $5(-1.2x) = (5 \cdot (-1.2))x = -6x$

115. $-(c + 7) + 2(c - 3) = -c - 7 + 2c - 6$
     $= c - 13$

117. $a^3 + 2a^2 + 4a - 2a^2 - 4a - 8 = a^3 - 8$

APPLICATIONS

119. $x + x + \cdots + x = 12x$

121. $x + (x + 4) + x + (x + 4) = (4x + 8)$ ft

WRITING

123. The distributive property only applies to multiple terms inside the parentheses. $3x$ is a single term.

REVIEW

125.
$$\frac{x-y^2}{2y-1+x} = \frac{-3-(-5)^2}{2(-5)-1+(-3)}$$
$$= \frac{-3-25}{-10-1-3} = \frac{-28}{-14} = 2$$

8.3: Solving Equations Using Properties of Equality

VOCABULARY

1. An <u>equation</u>, such as $x+1=7$, is a statement indicating that two expressions are equal.

3. To <u>solve</u> and equation means to find all values of the variable that make the equation true.

5. Equations with the same solutions are called <u>equivalent</u> equations.

CONCEPTS

7. a. $x+6$

   b. neither

   c. $5+6 \ne 12$, so no

   d. $6+6=12$, so yes

9. a. $a+c=b+c$ and $a-c=b-c$

   b. $ca=cb$ and $\dfrac{a}{c}=\dfrac{b}{c}$

11. a. $x-7+7=x$

    b. $y-2+2=y$

    c. $\dfrac{5t}{5}=t$

    d. $6 \cdot \dfrac{h}{6}=h$

NOTATION

13. $x-5=45$
    $x-5+5=45+5$
    $x=50$

Check:
$x-5=45$
$50-5 \stackrel{?}{=} 45$
$45=45$

True

15. a. is possibly equal to

    b. yes

GUIDED PRACTICE

17. $6+12 \stackrel{?}{=} 28$
    $18 = 28$

    6 is not a solution.

19. $2(-8)+3 \stackrel{?}{=} -15$
    $-16+3 \stackrel{?}{=} -15$
    $-13 = -15$

    -8 is not a solution.

21. $0.5(5) \stackrel{?}{=} 2.9$
    $2.5 = 2.9$

    5 is not a solution.

23. $33 - \dfrac{(-6)}{2} \stackrel{?}{=} 30$
    $33-(-3) \stackrel{?}{=} 30$
    $33+3 \stackrel{?}{=} 30$
    $36 = 30$

    -6 is not a solution.

25. $|-2-8| \stackrel{?}{=} 10$
    $|-10| \stackrel{?}{=} 10$
    $10 = 10$

    -2 is a solution.

27.
$$3(12) - 2 \stackrel{?}{=} 4(12) - 5$$
$$36 - 2 \stackrel{?}{=} 48 - 5$$
$$34 = 43$$

12 is not a solution.

29.
$$(-3)^2 - (-3) - 6 \stackrel{?}{=} 0$$
$$9 + 3 - 6 \stackrel{?}{=} 0$$
$$6 = 0$$

-3 is not a solution.

31.
$$\frac{2}{1+1} + 5 \stackrel{?}{=} \frac{12}{1+1}$$
$$\frac{2}{2} + 5 \stackrel{?}{=} \frac{12}{2}$$
$$1 + 5 \stackrel{?}{=} 6$$
$$6 = 6$$

1 is a solution.

33.
$$\frac{3}{4} - \frac{1}{8} \stackrel{?}{=} \frac{5}{8}$$
$$\frac{6}{8} - \frac{1}{8} \stackrel{?}{=} \frac{5}{8}$$
$$\frac{5}{8} = \frac{5}{8}$$

$\frac{3}{4}$ is a solution.

35.
$$(-3 - 4)(-3 + 3) \stackrel{?}{=} 0$$
$$(-7)(0) \stackrel{?}{=} 0$$
$$0 = 0$$

-3 is a solution.

37.
$$a - 5 = 66$$
$$a - 5 + 5 = 66 + 5$$
$$a = 71$$

Check:
$$71 - 5 \stackrel{?}{=} 66$$
$$66 = 66$$

39.
$$9 = p - 9$$
$$9 + 9 = p - 9 + 9$$
$$18 = p$$

Check:
$$9 \stackrel{?}{=} 18 - 9$$
$$9 = 9$$

41.
$$x - 1.6 = -2.5$$
$$x - 1.6 + 1.6 = -2.5 + 1.6$$
$$x = -0.9$$

Check:
$$-0.9 - 1.6 \stackrel{?}{=} -2.5$$
$$-2.5 = -2.5$$

43.
$$-3 + a = 0$$
$$-3 + a + 3 = 0 + 3$$
$$a = 3$$

Check:
$$-3 + 3 \stackrel{?}{=} 0$$
$$0 = 0$$

45.
$$d - \frac{1}{9} = \frac{7}{9}$$
$$d - \frac{1}{9} + \frac{1}{9} = \frac{7}{9} + \frac{1}{9}$$
$$d = \frac{8}{9}$$

Check:
$$\frac{8}{9} - \frac{1}{9} \stackrel{?}{=} \frac{7}{9}$$
$$\frac{7}{9} = \frac{7}{9}$$

47.
$x + 7 = 10$
$x + 7 - 7 = 10 - 7$
$x = 3$

Check:

$3 + 7 \stackrel{?}{=} 10$
$10 = 10$

49.
$s + \dfrac{1}{5} = \dfrac{4}{25}$
$s + \dfrac{1}{5} - \dfrac{1}{5} = \dfrac{4}{25} - \dfrac{1}{5}$
$s = \dfrac{4}{25} - \dfrac{5}{25}$
$s = -\dfrac{1}{25}$

Check:

$-\dfrac{1}{25} + \dfrac{1}{5} \stackrel{?}{=} \dfrac{4}{25}$
$-\dfrac{1}{25} + \dfrac{5}{25} \stackrel{?}{=} \dfrac{4}{25}$
$\dfrac{4}{25} = \dfrac{4}{25}$

51.
$3.5 + f = 1.2$
$3.5 + f - 3.5 = 1.2 - 3.5$
$f = -2.3$

Check:

$3.5 + (-2.3) \stackrel{?}{=} 1.2$
$3.5 - 2.3 \stackrel{?}{=} 1.2$
$1.2 = 1.2$

53.
$\dfrac{x}{15} = 3$
$15 \cdot \dfrac{x}{15} = 15 \cdot 3$
$x = 45$

Check:

$\dfrac{45}{15} \stackrel{?}{=} 3$
$3 = 3$

55.
$0 = \dfrac{v}{11}$
$0 \cdot 11 = \dfrac{v}{11} \cdot 11$
$0 = v$

Check: $0 \stackrel{?}{=} \dfrac{0}{11}$
$0 = 0$

57.
$\dfrac{d}{-7} = -3$
$-7 \cdot \dfrac{d}{-7} = -7(-3)$
$d = 21$

Check:

$\dfrac{21}{-7} \stackrel{?}{=} -3$
$-3 = -3$

59.
$\dfrac{y}{0.6} = -4.4$
$0.6 \cdot \dfrac{y}{0.6} = 0.6(-4.4)$
$y = -2.64$

Check:

$\dfrac{-2.64}{0.6} \stackrel{?}{=} -4.4$
$-4.4 = -4.4$

61.
$\dfrac{4}{5}t = 16$
$\dfrac{5}{4} \cdot \dfrac{4}{5}t = \dfrac{5}{4} \cdot 16$
$1t = \dfrac{80}{4}$
$t = 20$

63.
$$\frac{2}{3}c = 10$$
$$\frac{3}{2} \cdot \frac{2}{3}c = \frac{3}{2} \cdot 10$$
$$1c = \frac{3}{\cancel{2}} \cdot \cancel{10}^{5}$$
$$c = 15$$

65.
$$-\frac{7}{2}r = 21$$
$$\left(-\frac{2}{7}\right)\left(-\frac{7}{2}r\right) = -\frac{2}{7} \cdot 21$$
$$1r = -\frac{2}{\cancel{7}} \cdot \cancel{21}^{3}$$
$$r = -6$$

67.
$$-\frac{5}{4}h = -5$$
$$\left(-\frac{4}{5}\right)\left(-\frac{5}{4}h\right) = -\frac{4}{5}(-5)$$
$$1h = \frac{4}{\cancel{5}}(\cancel{5})$$
$$h = 4$$

69.
$$4x = 16$$
$$\frac{4x}{4} = \frac{16}{4}$$
$$x = 4$$

71.
$$63 = 9c$$
$$\frac{63}{9} = \frac{9c}{9}$$
$$7 = c$$

73.
$$23b = 23$$
$$\frac{23b}{23} = \frac{23}{23}$$
$$b = 1$$

75.
$$-8h = 48$$
$$\frac{-8h}{-8} = \frac{48}{-8}$$
$$h = -6$$

77.
$$-100 = -5g$$
$$\frac{-100}{-5} = \frac{-5g}{-5}$$
$$20 = g$$

79.
$$-3.4y = -1.7$$
$$\frac{-3.4y}{-3.4} = \frac{-1.7}{-3.4}$$
$$y = \frac{17}{34} = \frac{1}{2}$$
$$y = 0.5$$

81.
$$-x = 18$$
$$\frac{-x}{-1} = \frac{18}{-1}$$
$$x = -18$$

83.
$$-n = \frac{4}{21}$$
$$-1(-n) = -1\left(\frac{4}{21}\right)$$
$$n = -\frac{4}{21}$$

TRY IT YOURSELF

85.
$$8.9 = -4.1 + t$$
$$8.9 + 4.1 = -4.1 + t + 4.1$$
$$13 = t$$

Check:
$$8.9 \stackrel{?}{=} -4.1 + 13$$
$$8.9 = 8.9$$

87.
$$-2.5 = -m$$
$$\frac{-2.5}{-1} = \frac{-m}{-1}$$
$$2.5 = m$$

Check:
$$-2.5 = -2.5$$

89.
$$-\frac{9}{8}x = 3$$
$$\left(-\frac{8}{9}\right)\left(-\frac{9}{8}x\right) = \left(-\frac{8}{9}\right)3$$
$$1x = \left(-\frac{8}{\cancel{9}}\right)\cancel{3}$$
$$x = -\frac{8}{3}$$

Check:
$$-\frac{9}{8} \cdot \left(-\frac{8}{3}\right) \stackrel{?}{=} 3$$
$$\frac{\cancel{3} \cdot 3 \cdot \cancel{8}}{\cancel{8} \cdot \cancel{3}} \stackrel{?}{=} 3$$
$$3 = 3$$

91.
$$\frac{3}{4} = d + \frac{1}{10}$$
$$\frac{3}{4} - \frac{1}{10} = d + \frac{1}{10} - \frac{1}{10}$$
$$\frac{15}{20} - \frac{2}{20} = d$$
$$\frac{13}{20} = d$$

Check:
$$\frac{3}{4} \stackrel{?}{=} \frac{13}{20} + \frac{1}{10}$$
$$\frac{3}{4} \stackrel{?}{=} \frac{13}{20} + \frac{2}{20}$$
$$\frac{3}{4} \stackrel{?}{=} \frac{15}{20}$$
$$\frac{3}{4} = \frac{3}{4}$$

93.
$$-15x = -60$$
$$\frac{-15x}{-15} = \frac{-60}{-15}$$
$$x = 4$$

Check:
$$-15 \cdot 4 \stackrel{?}{=} -60$$
$$-60 = -60$$

95.
$$-10 = n - 5$$
$$-10 + 5 = n - 5 + 5$$
$$-5 = n$$

Check:
$$-10 \stackrel{?}{=} -5 - 5$$
$$-10 = -10$$

97.
$$\frac{h}{-40} = 5$$
$$-40\left(\frac{h}{-40}\right) = -40 \cdot 5$$
$$h = -200$$

Check:
$$\frac{-200}{-40} \stackrel{?}{=} 5$$
$$5 = 5$$

99.
$$a - 93 = 2$$
$$a - 93 + 93 = 2 + 93$$
$$a = 95$$

Check:
$$95 - 93 \stackrel{?}{=} 2$$
$$2 = 2$$

APPLICATIONS

101.
$$x + 115 = 180$$
$$x + 115 - 115 = 180 - 115$$
$$x = 65°$$

103.
$$\frac{x}{16} = 375,000$$
$$16 \cdot \frac{x}{16} = 16 \cdot 375,00$$
$$x = \$6,000,000$$

WRITING

105. It means to find all values of the variable that make the statement true.

107. The person neglected to subtract 2 from the right hand side of the equation.

REVIEW

109. $-9-3x = -9-3(-3) = -9+9 = 0$

111. $45-x$

8.4: More about Solving Equations

VOCABULARY

1. To solve and equation means to find all values of the variable that make the equation true.

3. When solving equations, simplify the expressions that make up the left and right sides of the equation before using the properties of equality to isolate the variable.

CONCEPTS

5. On the left side of the equation $4x+9=25$, the variable x is multiplied by 4 and then 9 is added to that product.

7. To solve $3x-5=1$, we first undo the subtraction of 5 by adding 5 to both sides. Then wee undo the multiplication of 3 by dividing both sides by 3.

9. a. $-2x-8=-24$

b. $-20 = 4(3x-4)-9x$
   $= 12x-16-9x = 3x-16$

11.
$$6(-2)+5 \stackrel{?}{=} 7$$
a. $-12+5 \stackrel{?}{=} 7$
$$-7 \stackrel{?}{=} 7$$
not a solution

$$8(-2+3) \stackrel{?}{=} 8$$
b. $8(1) \stackrel{?}{=} 8$
$$8 = 8$$
solution

NOTATION

13.
$$2x-7 = 21$$
$$2x-7+7 = 21+7$$
$$2x = 28$$
$$\frac{2x}{2} = \frac{28}{2}$$
$$x = 14$$

$$2x-7 = 21$$
$$2(14)-7 \stackrel{?}{=} 21$$
$$28-7 \stackrel{?}{=} 21$$
$$21 = 21$$

14 is the solution.

GUIDED PRACTICE

15.
$$2x+5 = 17$$
$$2x+5-5 = 17-5$$
$$2x = 12$$
$$\frac{2x}{2} = \frac{12}{2}$$
$$x = 6$$

17.
$$5q-2 = 23$$
$$5q-2+2 = 23+2$$
$$5q = 25$$
$$\frac{5q}{5} = \frac{25}{5}$$
$$q = 5$$

19.
$$-33 = 5t+2$$
$$-33-2 = 5t+2-2$$
$$-35 = 5t$$
$$\frac{-35}{5} = \frac{5t}{5}$$
$$-7 = t$$

21. $0.7 + 4y = 1.7$
$0.7 + 4y - 0.7 = 1.7 - 0.7$
$4y = 1$
$\dfrac{4y}{4} = \dfrac{1}{4}$
$y = 0.25$

23. $-5 - 2d = 0$
$-5 - 2d + 5 = 0 + 5$
$-2d = 5$
$\dfrac{-2d}{-2} = \dfrac{5}{-2}$
$d = -\dfrac{5}{2}$

25. $12 = -7a - 9$
$12 + 9 = -7a - 9 + 9$
$21 = -7a$
$\dfrac{21}{-7} = \dfrac{-7a}{-7}$
$-3 = a$

27. $-3 = -3p + 7$
$-3 - 7 = -3p + 7 - 7$
$-10 = -3p$
$\dfrac{-10}{-3} = \dfrac{-3p}{-3}$
$\dfrac{10}{3} = p$

29. $\dfrac{2}{3}t + 2 = 6$
$\dfrac{2}{3}t + 2 - 2 = 6 - 2$
$\dfrac{2}{3}t = 4$
$\dfrac{3}{2} \cdot \dfrac{2}{3}t = \dfrac{3}{2} \cdot 4$
$t = \dfrac{12}{2}$
$t = 6$

31. $\dfrac{5}{6}k - 5 = 10$
$\dfrac{5}{6}k - 5 + 5 = 10 + 5$
$\dfrac{5}{6}k = 15$
$\dfrac{6}{5} \cdot \dfrac{5}{6}k = \dfrac{6}{5} \cdot 15$
$1k = \dfrac{6}{\cancel{5}} \cdot \cancel{15}^{3}$
$k = 18$

33. $-\dfrac{7}{16}h + 28 = 21$
$-\dfrac{7}{16}h + 28 - 28 = 21 - 28$
$-\dfrac{7}{16}h = -7$
$\left(-\dfrac{16}{7}\right)\left(-\dfrac{7}{16}h\right) = \left(-\dfrac{16}{7}\right)(-7)$
$1h = \dfrac{16}{\cancel{7}} \cdot \cancel{7}$
$h = 16$

35. $-1.7 = 1.2 - x$
$-1.7 - 1.2 = 1.2 - x - 1.2$
$-2.9 = -x$
$\dfrac{-2.9}{-1} = \dfrac{-x}{-1}$
$2.9 = x$

37. $-6 - y = -2$
$-6 - y + 6 = -2 + 6$
$-y = 4$
$\dfrac{-y}{-1} = \dfrac{4}{-1}$
$y = -4$

39.
$3(2y-2)-y=5$
$6y-6-y=5$
$5y-6=5$
$5y-6+6=5+6$
$5y=11$
$\dfrac{5y}{5}=\dfrac{11}{5}$

41.
$9(x+11)+5(13-x)=0$
$9x+99+65-5x=0$
$4x+164=0$
$4x+164-164=0-164$
$4x=-164$
$\dfrac{4x}{4}=\dfrac{-164}{4}$
$x=-41$

43.
$-(4-m)=-10$
$-4+m=-10$
$-4+m+4=-10+4$
$m=-6$

45.
$10.08=4(0.5x+2.5)$
$10.08=2x+10$
$10.08-10=2x+10-10$
$0.08=2x$
$\dfrac{0.08}{2}=\dfrac{2x}{2}$
$0.04=x$

47.
$6a-3(3a-4)=30$
$6a-9a+12=30$
$-3a+12=30$
$-3a+12-12=30-12$
$-3a=18$
$\dfrac{-3a}{-3}=\dfrac{18}{-3}$
$a=-6$

49.
$-(19-3s)-(8s+1)=35$
$-19+3s-8s-1=35$
$-5s-20=35$
$-5s-20+20=35+20$
$-5s=55$
$\dfrac{-5s}{-5}=\dfrac{55}{-5}$
$s=-11$

51.
$5x=4x+7$
$5x-4x=4x+7-4x$
$x=7$

53.
$8y+44=4y$
$8y+44-8y=4y-8y$
$44=-4y$
$\dfrac{44}{-4}=\dfrac{-4y}{-4}$
$-11=y$

55.
$60r-50=15r-5$
$60r-50-15r=15r-5-15r$
$45r-50=-5$
$45r-50+50=-5+50$
$45r=45$
$\dfrac{45r}{45}=\dfrac{45}{45}$
$r=1$

57.
$8y-2=4y+16$
$8y-2+2=4y+16+2$
$8y=4y+18$
$8y-4y=4y+18-4y$
$4y=18$
$\dfrac{4y}{4}=\dfrac{18}{4}$
$y=\dfrac{9}{2}$

59.
$3(A+2)+4A = 2(A-7)$
$3A+6+4A = 2A-14$
$7A+6 = 2A-14$
$7A+6-6 = 2A-14-6$
$7A = 2A-20$
$7A-2A = 2A-20-2A$
$5A = -20$
$\dfrac{5A}{5} = \dfrac{-20}{5}$
$A = -4$

61.
$2-3(x-5) = 4(x-1)$
$2-3x+15 = 4x-4$
$17-3x = 4x-4$
$17-3x+3x = 4x-4+3x$
$17 = 7x-4$
$17+4 = 7x-4+4$
$21 = 7x$
$\dfrac{21}{7} = \dfrac{7x}{7}$
$3 = x$

TRY IT YOURSELF

63.
$3x-8-4x-7x = -2-8$
$-8x-8 = -10$
$-8x-8+8 = -10+8$
$-8x = -2$
$\dfrac{-8x}{-8} = \dfrac{-2}{-8}$
$x = \dfrac{1}{4}$

65.
$4(d-5)+20 = 5-2d$
$4d-20+20 = 5-2d$
$4d = 5-2d$
$4d+2d = 5-2d+2d$
$6d = 5$
$\dfrac{6d}{6} = \dfrac{5}{6}$

Chapter 8 An Introduction to Algebra 171

67.
$30x-12 = 1{,}338$
$30x-12+12 = 1{,}338+12$
$30x = 1{,}350$
$\dfrac{30x}{30} = \dfrac{1{,}350}{30}$
$x = 45$

69.
$-7 = \dfrac{3}{7}r+14$
$-7-14 = \dfrac{3}{7}r+14-14$
$-21 = \dfrac{3}{7}r$
$\dfrac{7}{3}(-21) = \dfrac{7}{3} \cdot \dfrac{3}{7}r$
$\dfrac{7}{\cancel{3}}\left(\cancel{-21}\right) = 1r$
$-49 = r$

71.
$10-2y = 8$
$10-2y-10 = 8-10$
$-2y = -2$
$\dfrac{-2y}{-2} = \dfrac{-2}{-2}$
$y = 1$

73.
$9+5(r+3) = 6+3(r-2)$
$9+5r+15 = 6+3r-6$
$24+5r = 3r$
$24+5r-5r = 3r-5r$
$24 = -2r$
$\dfrac{24}{-2} = \dfrac{-2r}{-2}$
$-12 = r$

75.
$$-\frac{2}{3}z + 4 = 8$$
$$-\frac{2}{3}z + 4 - 4 = 8 - 4$$
$$-\frac{2}{3}z = 4$$
$$\left(-\frac{3}{2}\right)\left(-\frac{2}{3}z\right) = \left(-\frac{3}{2}\right)4$$
$$1z = \frac{-12}{2}$$
$$z = -6$$

77.
$$-2(9-3s)-(5s+2) = -25$$
$$-18 + 6s - 5s - 2 = -25$$
$$s - 20 = -25$$
$$s - 20 + 20 = -25 + 20$$
$$s = -5$$

79.
$$9a - 2.4 = 7a + 4.6$$
$$9a - 2.4 - 7a = 7a + 4.6 - 7a$$
$$2a - 2.4 = 4.6$$
$$2a - 2.4 + 2.4 = 4.6 + 2.4$$
$$2a = 7$$
$$a = \frac{7}{2} = 3.5$$

WRITING

81. Yes, they will get the same answer since they are doing the same thing to both sides of the equation.

One student will get $-\frac{5}{2} = x$ and the other will get

$x = -\frac{5}{2}$ - they are equivalent.

REVIEW

83. commutative property of multiplication

85. associative property of addition

8.5: Using Equations to Solve Application Problems

VOCABULARY

1. The five-step problem-solving strategy is:

   *<u>Analyze</u> the problem

   *Form an <u>equation</u>

   *<u>Solve</u> the equation

   *State the <u>conclusion</u>

   *<u>Check</u> the result

3. Phrases such as *distributed equally* and *sectioned off uniformly* indicate the operation of <u>division.</u>

5. Words such as *extended* and *reclaimed* indicated the operation of <u>addition.</u>

CONCEPTS

7. Key word: <u>borrow</u>

   Translation: <u>add</u>

9. Key word: <u>equal-size discussion groups</u>

   Translation: <u>division</u>

11. $s + 6$

13. $g - 100$

15. **Analyze**

   * The scroll is <u>1,700</u> years old.

   * The scroll is <u>425</u> years older than the jar.

   * How old is the <u>jar</u>?

   Form Let x = the <u>age</u> of the jar. Now

   Key phrase: older than

   Translation: <u>addition</u>

   Scroll's age is 425 years plus jar's age

      1,700   = 425   + x

Solve

$1,700 = 425 + x$

$1,700 - 425 = 425 + x - 425$

$1,275 = x$

State The jar is 1,275 years old.

Check
$$\begin{array}{r} 1,275 \\ + 425 \\ \hline 1,700 \end{array}$$

17. **Analyze**

    * There are 88 seats on the plane.

    * There are 10 times as may economy as first-class seats.

    *Find the number of first-class seats and the number of economy seats.

    Form Let x = the number of first-class seats. Now

    Key phrase: ten times as many

    Translation: multiply by 10.

    # first-class seats plus # economy seats is 88.

    $\quad\quad x \quad\quad + \quad\quad 10x \quad\quad = 88$

    Solve

    $x + 10x = 88$

    $11x = 88$

    $\dfrac{11x}{11} = \dfrac{88}{11}$

    $x = 8$

    State There are 8 first class seats and 80 economy seats.

    Check The number of economy seats, 80, is 10 times the number of first-class seats, 8. Also, if we add the numbers of seats, we get:

    $$\begin{array}{r} 80 \\ +8 \\ \hline 88 \end{array}$$

Chapter 8 An Introduction to Algebra 173

GUIDED PRACTICE

19. Let x = the amount of money she needs to borrow.

    $68,500 + x = 316,500$

    $68,500 + x - 68,500 = 316,500 - 68,500$

    $x = 248,000$

    She needs to borrow $248,000

21. Let x = the number of words per minute she could read prior to the course.

    $3x = 399$

    $\dfrac{3x}{3} = \dfrac{399}{3}$

    $x = 133$

    Alicia could read 133 words per minute before the course.

23. Let x = the number of months to reach his goal.

    $15 + 5x = 100$

    $15 + 5x - 15 = 100 - 15$

    $5x = 85$

    $\dfrac{5x}{5} = \dfrac{85}{5}$

    $x = 17$

    It will take the salesman 17 months to reach his goal.

25. Let x = the number of scholarships awarded last year.

    $x + (x + 6) = 20$

    $2x + 6 = 20$

    $2x + 6 - 6 = 20 - 6$

    $2x = 14$

    $\dfrac{2x}{2} = \dfrac{14}{2}$

    $x = 7$

    Last year there were 7 scholarships awarded. This year there were 7 + 6 = 13 scholarships awarded.

APPLICATIONS

27. Let x = the number of payments she has made.

$$600 - 30x = 420$$
$$600 - 30x - 600 = 420 - 600$$
$$-30x = -180$$
$$\frac{-30x}{-30} = \frac{-180}{-30}$$
$$x = 6$$

She has made 6 payments so far.

29. Let x = the amount 50 Cent earned in 2008.

$$x - 82 = 68$$
$$x - 82 + 82 = 68 + 82$$
$$x = 150$$

50 Cent earned $150 million in 2008.

31. Let w = the width of the room. Then the length is $2w$.

$$P = 2l + 2w$$
$$60 = 2 \cdot 2w + 2w$$
$$60 = 4w + 2w$$
$$60 = 6w$$
$$\frac{60}{6} = \frac{6w}{6}$$
$$10 = w$$

The room's width is 10 feet. The length is $2 \cdot 10 = 20$ feet.

33. Let w = the weight of the object on the moon.

$$6w = 330$$
$$\frac{6w}{6} = \frac{330}{6}$$
$$w = 55$$

The scale would register 55 pounds on the moon.

35. Let s = the number of scenes in the first act.

$$s + 6 + 5 + 5 + 3 = 24$$
$$s + 19 = 24$$
$$s + 19 - 19 = 24 - 19$$
$$s = 5$$

There were 5 scenes in the opening act.

37. Let x = the amount of money in benefits.

$$45 + x = 52$$
$$45 + x - 45 = 52 - 45$$
$$x = 7$$

The benefits package is valued at $7,000.

39. Let s = the score on the first game.

$$s + 9,485 = 11,053$$
$$s + 9,485 - 9,485 = 11,053 - 9,845$$
$$s = 1,568$$

His first *Sonic* game score was 1,568 points.

41. Let c = the amount of time allotted for the commercials.

$$c + (c + 18) = 30$$
$$2c + 18 = 30$$
$$2c + 18 - 18 = 30 - 18$$
$$2c = 12$$
$$\frac{2c}{2} = \frac{12}{2}$$
$$c = 6$$

The manager budgeted 6 minutes for the commercials and 6 + 18 = 24 minutes for the program.

43. Let l = the time spend in the lab. Then $l + 50$ is the time spent in the lecture.

$$l + (l + 50) = 250$$
$$2l + 50 = 250$$
$$2l + 50 - 50 = 250 - 50$$
$$2l = 200$$
$$\frac{2l}{2} = \frac{200}{2}$$
$$l = 100$$

The students spend 100 minutes in the lab and 100 + 50 = 150 minutes in the lecture.

45. Let s = the number of the shelter's calls

$$4 \cdot 8 = s$$
$$32 = s$$

The shelter received 32 calls per day after the news feature.

47. Let w = the length of his wait three days ago.

$$20 + 15 = w$$
$$35 = w$$

Three days ago he waited for 35 minutes.

49. Let x = the original estimate.

$$10x = 540$$
$$\frac{10x}{10} = \frac{540}{10}$$
$$x = 54$$

The original estimate was $54 million.

51. Let r = the monthly rent on the apartment.

$$3(425 - 100) = r$$
$$3(325) = r$$
$$975 = r$$

The monthly rent on the apartment was $975

53. Let x = the number of training sessions still needed.

$$24 + 6x = 48$$
$$24 + 6x - 24 = 48 - 24$$
$$6x = 24$$
$$\frac{6x}{6} = \frac{24}{6}$$
$$x = 4$$

She still has 4 sessions left before she is certified.

WRITING

55. Checking my work is the most difficult, since I don't like to think I made any mistakes.

57. It means to rewrite a sentence as an equation.

59. 'Age of father' + 'age of son' is 50

REVIEW

61. $100 = 2 \cdot 2 \cdot 5 \cdot 5$
$120 = 2 \cdot 2 \cdot 2 \cdot 3 \cdot 5$
$LCM(100, 120) = 2 \cdot 2 \cdot 2 \cdot 3 \cdot 5 \cdot 5 = 600$
$GCF(100, 120) = 2 \cdot 2 \cdot 5 = 20$

63. $14 = 2 \cdot 7$
$140 = 2 \cdot 2 \cdot 5 \cdot 7$
$LCM(14, 140) = 2 \cdot 2 \cdot 5 \cdot 7 = 140$
$GCF(14, 140) = 2 \cdot 7 = 14$

65. $8 = 2 \cdot 2 \cdot 2$
$9 = 3 \cdot 3$
$49 = 7 \cdot 7$
$LCM(8, 9, 49) = 2 \cdot 2 \cdot 2 \cdot 3 \cdot 3 \cdot 7 \cdot 7 = 3{,}528$
$GCF(8, 9, 49) = 1$

67. $66 = 2 \cdot 3 \cdot 11$
$198 = 2 \cdot 3 \cdot 3 \cdot 11$
$242 = 2 \cdot 11 \cdot 11$
$LCM(66, 198, 242) = 2 \cdot 3 \cdot 3 \cdot 11 \cdot 11 = 2{,}178$
$GCF(66, 198, 242) = 2 \cdot 11 = 22$

8.6: Multiplication Rules for Exponents

VOCABULARY

1. Expressions such as $x^4, 10^3,$ and $(5t)^2$ are called <u>exponential</u> expressions.

CONCEPTS

3. a. $(3x)^4 = 3x \cdot 3x \cdot 3x \cdot 3x$

   b. $(-5y)(-5y)(-5y) = (-5y)^3$

5. a. add

   b. multiply

   c. multiply

7. a. $x^2 + x^2 = 2x^2$ b. $x^2 \cdot x^2 = x^{2+2} = x^4$

9. a. doesn't simplify b. $x^3 \cdot x^2 = x^{3+2} = x^5$

NOTATION

11. $(x^4 x^2)^3 = (x^6)^3$
    $= x^{18}$

GUIDED PRACTICE

13. base: 4 ; exponent 2

15. base x ; exponent 5

17. base $-3x$; exponent 2

19. base y ; exponent 6

21. base m ; exponent 12

23. base $y+9$; exponent 4

25. m^5

27. $(4t)^4$

29. $4t^5$

31. $a^2 b^3$

33. $5^3 \cdot 5^4 = 5^{3+4} = 5^7$

35. $a^3 \cdot a^3 = a^{3+3} = a^6$

37. $bb^2 b^3 = b^{1+2+3} = b^6$

39. $(c^5)(c^8) = c^{5+8} = c^{13}$

41. $(a^2 b^3)(a^3 b^3) = a^2 a^3 b^3 b^3$
    $= a^{2+3} b^{3+3} = a^5 b^6$

43. $cd^4 \cdot cd = ccd^4 d = c^{1+1} d^{4+1} = c^2 d^5$

45. $x^2 \cdot y \cdot x \cdot y^{10} = x^{2+1} y^{1+10} = x^3 y^{11}$

47. $m^{100} \cdot m^{100} = m^{100+100} = m^{200}$

49. $(3^2)^4 = 3^{2 \cdot 4} = 3^8$

51. $[(-4.3)^3]^8 = (-4.3)^{3 \cdot 8} = (-4.3)^{24}$

53. $(m^{50})^{10} = m^{50 \cdot 10} = m^{500}$

55. $(y^5)^3 = y^{5 \cdot 3} = y^{15}$

57. $(x^2 x^3)^5 = (x^5)^5 = x^{25}$

59. $(p^2 p^3)^5 = (p^5)^5 = p^{25}$

61. $(t^3)^4 (t^2)^3 = t^{12} t^6 = t^{18}$

63. $(u^4)^2 (u^3)^2 = u^8 u^6 = u^{14}$

65. $(6a)^2 = 6^2 a^2 = 36 a^2$

67. $(5y)^4 = 5^4 y^4 = 625 y^4$

69. $(3a^4 b^7)^3 = 3^3 (a^4)^3 (b^7)^3 = 27 a^{12} b^{21}$

71. $(-2r^2 s^3)^3 = (-2)^3 (r^2)^3 (s^3)^3 = -8 r^6 s^9$

73. $(2c^3)^3(3c^4)^2 = 2^3(c^3)^3 \cdot 3^2(c^4)^2$
$= 8c^9 \cdot 9c^8 = 8 \cdot 9 \cdot c^9 \cdot c^8 = 72c^{17}$

75. $(10d^7)^2(4d^9)^3 = 10^2(d^7)^2 \cdot 4^3(d^9)^3$
$= 100d^{14} \cdot 64d^{27} = 100 \cdot 64 \cdot d^{14} \cdot d^{27}$
$= 6,400d^{41}$

TRY IT YOURSELF

77. $(7a^9)^2 = 7^2(a^9)^2 = 49a^{18}$

79. $t^4 \cdot t^5 \cdot t = t^{4+5+1} = t^{10}$

81. $y^3 y^2 y^4 = y^{3+2+4} = y^9$

83. $(-6a^3b^2)^3 = (-6)^3(a^3)^3(b^2)^3 = -216a^9b^6$

85. $(n^4n)^3(n^3)^6 = (n^5)^3(n^3)^6 = n^{15}n^{18} = n^{33}$

87. $(b^2b^3)^{12} = (b^5)^{12} = b^{60}$

89. $(2b^4b)^5(3b)^2 = (2b^5)^5(3b)^2$
$= 2^5(b^5)^5 3^2 b^2 = 32b^{25} \cdot 9b^2$
$= 32 \cdot 9 \cdot b^{25} \cdot b^2 = 288b^{27}$

91. $(c^2)^3(c^4)^2 = c^6 \cdot c^8 = c^{14}$

93. $(3s^4t^3)^3(2st)^4 = 3^3(s^4)^3(t^3)^3 \cdot 2^4 s^4 t^4$
$= 27s^{12}t^9 \cdot 16s^4t^4 = 27 \cdot 16 \cdot s^{12} \cdot s^4 \cdot t^9 \cdot t^4$
$= 432s^{16}t^{13}$

95. $x \cdot x^2 \cdot x^3 \cdot x^4 \cdot x^5 = x^{1+2+3+4+5} = x^{15}$

APPLICATIONS

97. $(5x)^2 = 5^2 x^2 = 25x^2 \text{ ft}^2$

WRITING

99. When multiplying two expressions with the same base, keep the original base.

REVIEW

101. $\dfrac{18}{24} = \dfrac{3 \cdot \cancel{6}}{4 \cdot \cancel{6}} = \dfrac{3}{4}$

103. $\dfrac{-25}{-5} = \dfrac{5 \cdot \cancel{5}}{\cancel{5}} = 5$

105. $2\left(\dfrac{12}{-3}\right) + 3(5) = 2(-4) + 15$
$= -8 + 15 = 7$

107. $-x = -12$
$\dfrac{-x}{-1} = \dfrac{-12}{-1}$
$x = 12$

Chapter 8 Review

1. a. $6 \cdot b = 6b$

   b. $x \cdot y \cdot z = xyz$

   c. $2(t) = 2t$

3. a. factor b. term

5. a. 16, -1, 25 b. $\dfrac{1}{2}, 1$

7. a. $h + 25$

   b. $100 - 2s$

   c. $\dfrac{1}{2}t - 6$

   d. $|2 - a^2|$

9. a. $(x+1)$ in

   b. $\dfrac{p}{8}$ lb

11. $2x^2 + 3x + 7$
$2(5)^2 + 3(5) + 7$
$= 2 \cdot 25 + 15 + 7$
$= 50 + 15 + 7$
$= 72$

13. $b^2 - 4ac$
$(-10)^2 - 4(3)(5)$
$= 100 - 60 = 40$

15. $4(7w) = (4 \cdot 7)w = 28w$

17. $0.4(5.2f) = (0.4 \cdot 5.2)f = 2.08f$

19. $5(x+3) = 5 \cdot x + 5 \cdot 3 = 5x + 15$

21. $\dfrac{3}{4}(4c - 8) = \dfrac{3}{4} \cdot 4c - \dfrac{3}{4} \cdot 8 = 3c - 6$

23. $7a, 9a$

25. $8p + 5p - 4p = 13p - 4p = 9p$

27. $n + n + n + n = 2n + 2n = 4n$

29. $55.7k^2 - 55.6k^2 = 0.1k^2$

31. does not simplify

33. $\dfrac{3}{5}w - \left(-\dfrac{2}{5}w\right) = \dfrac{3}{5}w + \dfrac{2}{5}w$
$= \dfrac{5}{5}w = 1w = w$

35. a. $1x = x$ b. $-1x = -x$
c. $4x - (-1) = 4x + 1$ d. $4x + (-1) = 4x - 1$

37. $x - 34 = 50$
$84 - 34 \stackrel{?}{=} 50$
$50 = 50$

84 is a solution.

39. $\dfrac{x}{5} = 6$
$\dfrac{-30}{5} \stackrel{?}{=} 6$
$-6 = 6$

-30 is not a solution.

41. $5b - 2 = 3b - 8$
$5(-3) - 2 \stackrel{?}{=} 3(-3) - 8$
$-15 - 2 \stackrel{?}{=} -9 - 8$
$-17 = -17$

-3 is a solution.

43. An <u>equation</u> is a statement indicating that two expressions are equal.

45. $x - 9 = 12$
$x - 9 + 9 = 12 + 9$
$x = 21$

Check:
$x - 9 = 12$
$21 - 9 \stackrel{?}{=} 12$
$12 = 12$

The solution checks.

47. $a + 3.7 = -16.9$
$a + 3.7 - 3.7 = -16.9 - 3.7$
$a = -20.6$

Check:
$a + 3.7 = -16.9$
$-20.6 + 3.7 \stackrel{?}{=} -16.9$
$-16.9 = -16.9$

The solution checks.

49. $120 = 5c$
$\dfrac{120}{5} = \dfrac{5c}{5}$
$24 = c$

Check:

$120 = 5c$

$120 \stackrel{?}{=} 5 \cdot 24$

$120 = 120$

The solution checks.

51. $\dfrac{4}{3}t = -12$

$\dfrac{3}{4} \cdot \dfrac{4}{3}t = \dfrac{3}{4}(-12)$

$t = -9$

Check:

$\dfrac{4}{3}t = -12$

$\dfrac{4}{3}(-9) \stackrel{?}{=} -12$

$-12 = -12$

The solution checks.

53. $6b = 0$

$\dfrac{6b}{6} = \dfrac{0}{6}$

$b = 0$

Check:

$6b = 0$

$6(0) \stackrel{?}{=} 0$

$0 = 0$

The solution checks.

55. $5x + 4 = 14$

$5x + 4 - 4 = 14 - 4$

$5x = 10$

$\dfrac{5x}{5} = \dfrac{10}{5}$

$x = 2$

Check:

$5x + 4 = 14$

$5 \cdot 2 + 4 \stackrel{?}{=} 14$

$10 + 4 \stackrel{?}{=} 14$

$14 = 14$

The solution checks.

57. $\dfrac{n}{5} - 2 = 4$

$\dfrac{n}{5} - 2 + 2 = 4 + 2$

$\dfrac{n}{5} = 6$

$5\left(\dfrac{n}{5}\right) = 5(6)$

$n = 30$

Check:

$\dfrac{n}{5} - 2 = 4$

$\dfrac{30}{5} - 2 \stackrel{?}{=} 4$

$6 - 2 \stackrel{?}{=} 4$

$4 = 4$

The solution checks.

59. $12a - 9 = 4a + 15$

$12a - 9 + 9 = 4a + 15 + 9$

$12a = 4a + 24$

$12a - 4a = 4a + 24 - 4a$

$8a = 24$

$\dfrac{8a}{8} = \dfrac{24}{8}$

$a = 3$

Check:

$12a - 9 = 4a + 15$

$12 \cdot 3 - 9 \stackrel{?}{=} 4 \cdot 3 + 15$

$36 - 9 \stackrel{?}{=} 12 + 15$

$27 = 27$

The solution checks.

61. $5(2x - 4) - 5x = 0$

$10x - 20 - 5x = 0$

$5x - 20 = 0$

$5x - 20 + 20 = 0 + 20$

$5x = 20$

$\dfrac{5x}{5} = \dfrac{20}{5}$

$x = 4$

Check:

$5(2x - 4) - 5x = 0$

$5(2 \cdot 4 - 4) - 5 \cdot 4 \stackrel{?}{=} 0$

$5(4) - 20 \stackrel{?}{=} 0$

$20 - 20 \stackrel{?}{=} 0$

$0 = 0$

The solution checks.

63. $2(m + 40) - 6m = 3(4m - 80)$

$2m + 80 - 6m = 12m - 240$

$-4m + 80 = 12m - 240$

$-4m + 80 - 80 = 12m - 240 - 80$

$-4m = 12m - 320$

$-4m - 12m = 12m - 320 - 12m$

$-16m = -320$

$\dfrac{-16m}{-16} = \dfrac{-320}{-16}$

$m = 20$

Check:

$2(m + 40) - 6m = 3(4m - 80)$

$2(20 + 40) - 6(20) \stackrel{?}{=} 3(4(20) - 80)$

$2(60) - 120 \stackrel{?}{=} 3(80 - 80)$

$120 - 120 \stackrel{?}{=} 3(0)$

$0 = 0$

The solution checks.

65. Let x be the amount the need to borrow.

$25{,}000 + x = 122{,}750$

$25{,}000 + x - 25{,}000 = 122{,}750 - 25{,}000$

$x = 97{,}750$

They need to borrow $97,750.

67. Let x be the original cost estimate.

$3x = 81$

$\dfrac{3x}{3} = \dfrac{81}{3}$

$x = 27$

The original estimate was $27 million.

69. Let x be the number of hours to cool.

$71 - 7x = 29$

$71 - 7x - 71 = 29 - 71$

$-7x = -42$

$\dfrac{-7x}{-7} = \dfrac{-42}{-7}$

$x = 6$

It would take 6 hours to cool.

71. Let x be how many miles she walks.

$x + (x + 3) = 15$

$2x + 3 = 15$

$2x + 3 - 3 = 15 - 3$

$2x = 12$

$\dfrac{2x}{2} = \dfrac{12}{2}$

$x = 6$

She walks 6 miles, so she runs $6 + 3 = 9$ miles.

73. Let w be the width of the parking lot.

$2w + 2(4w) = 250$
$2w + 8w = 250$
$10w = 250$
$\dfrac{10w}{10} = \dfrac{250}{10}$
$w = 25$

The parking lot is 25 feet wide and $4 \cdot 25 = 100$ feet long.

75. a. base n, exponent 12

   b. base $2x$, exponent 6

   c. base r, exponent 4

   d. base $(y-7)$, exponent 3

77. a. $x^2 \cdot x^2 = x^{2+2} = x^4$

   b. $x^2 + x^2 = 2x^2$

   c. $x \cdot x^2 = x^{1+2} = x^3$

   d. does not simplify

79. $7^4 \cdot 7^8 = 7^{4+8} = 7^{12}$

81. $(y^7)^3 = y^{21}$

83. $(6^3)^{12} = 6^{36}$

85. $(-16s^3)^2 s^4 = (-16)^2 (s^3)^2 s^4$
   $= 256 s^6 s^4 = 256 s^{10}$

87. $\left[(-9)^3\right]^5 = (-9)^{15}$

89. $(2x^2 x^3)^3 = (2x^5)^3 = 2^3 (x^5)^3 = 8x^{15}$

91. $(3a^4)^2 (2a^3)^3 = 3^2 (a^4)^2 2^3 (a^3)^3$
   $= 9a^8 \cdot 8a^9 = 72 a^{17}$

93. $(4m^3)^3 (2m^2)^2 = 4^3 (m^3)^3 2^2 (m^2)^2$
   $= 64 m^9 \cdot 4 m^4 = 256 m^{13}$

Chapter 8 Test

1. a. <u>Variables</u> are letters (or symbols) that stand for numbers.

   b. To perform the multiplication $3(x+4)$ we use the <u>distributive</u> property.

   c. Terms such as $7x^2$ and $5x^2$, which have the same variables raised to exactly the same power, are called <u>like</u> terms.

   d. When we write $4x + x$ as $5x$, we say we have <u>combined</u> like terms.

   e. The coefficient of the term $9y$ is 9.

   f. To evaluate $y^2 + 9y - 3$ for $y = -5$ we <u>substitute</u> -5 for y and apply the order of operations rule.

   g. Variables and/or numbers can be combined with the operations of arithmetic to create algebraic <u>expressions</u>.

   h. An <u>equation</u> is a statement indicating that two expressions are equal.

   i. To <u>solve</u> an equation means to find all values of the variable that make the equation true.

   j. To <u>check</u> the solution of an equation, we substitute the value for the variable in the original equation and determine whether the result is a true statement.

3. $(s - 10)$ in

5. three-fourths of t

7. a. factor   b. term

9. $\dfrac{x - 16}{x}$

   $\dfrac{4 - 16}{4} = \dfrac{-12}{4} = -3$

11. a. $9 \cdot 4x = (9 \cdot 4)x = 36x$

   b. $-10(12t) = (-10 \cdot 12)t = -120t$

   c. $18\left(\dfrac{2}{3}x\right) = \left(18 \cdot \dfrac{2}{3}\right)x = 12x$

   d.
   $-4(-6)(-3m) = 24(-3m) = (24(-3))m = -72m$

13. $12m^2$, $2m^2$

15. $4(2y+3)-5(y+3)$
   $= 8y+12-5y-15$
   $= 3y-3$

17. $x+6=10$
   $x+6-6=10-6$
   $x=4$

   Check:

   $x+6=10$
   $4+6\stackrel{?}{=}10$
   $10=10$

   The result checks.

19. $5t=55$
   $\dfrac{5t}{5}=\dfrac{55}{5}$
   $t=11$

   Check:

   $5t=55$
   $5 \cdot 11\stackrel{?}{=}55$
   $55=55$

   The result checks.

21. $d-\dfrac{1}{3}=\dfrac{1}{6}$

   $d-\dfrac{1}{3}+\dfrac{1}{3}=\dfrac{1}{6}+\dfrac{1}{3}$

   $d=\dfrac{1}{6}+\dfrac{2}{6}=\dfrac{3}{6}=\dfrac{1}{2}$

   Check:

   $d-\dfrac{1}{3}=\dfrac{1}{6}$

   $\dfrac{1}{2}-\dfrac{1}{3}\stackrel{?}{=}\dfrac{1}{6}$

   $\dfrac{3}{6}-\dfrac{2}{6}\stackrel{?}{=}\dfrac{1}{6}$

   $\dfrac{1}{6}=\dfrac{1}{6}$

   The result checks.

23. $15a-10=20$
   $15a-10+10=20+10$
   $15a=30$
   $\dfrac{15a}{15}=\dfrac{30}{15}$
   $a=2$

   Check:

   $15a-10=20$
   $15 \cdot 2-10\stackrel{?}{=}20$
   $30-10\stackrel{?}{=}20$
   $20=20$

   The result checks.

25. $3.6-r=9.8$
   $3.6-r-3.6=9.8-3.6$
   $-r=6.2$
   $\dfrac{-r}{-1}=\dfrac{6.2}{-1}$
   $r=-6.2$

Check:

$3.6 - r = 9.8$

$3.6 - (-6.2) \stackrel{?}{=} 9.8$

$3.6 + 6.2 \stackrel{?}{=} 9.8$

$9.8 = 9.8$

The solution checks.

27. $-\dfrac{15}{16}x + 15 = 0$

$-\dfrac{15}{16}x + 15 - 15 = 0 - 15$

$-\dfrac{15}{16}x = -15$

$\left(-\dfrac{16}{15}\right)\left(-\dfrac{15}{16}x\right) = \left(-\dfrac{16}{15}\right)(-15)$

$x = 16$

Check:

$-\dfrac{15}{16}x + 15 = 0$

$\dfrac{-15}{16} \cdot 16 + 15 \stackrel{?}{=} 0$

$-15 + 15 \stackrel{?}{=} 0$

$0 = 0$

The result checks.

29. Let I = the intensity of the engine.

$81 + 29 = I$

$110 = I$

The intensity of the engine is 110 decibels.

31. Let s be the number of strings.

$19 + 23 + 2 + s = 98$

$44 + s = 98$

$44 + s - 44 = 98 - 44$

$s = 54$

There are 54 strings in the orchestra.

33. Let s be the smaller number.

$s + (s + 17) = 63$

$2s + 17 = 63$

$2s = 63 - 17$

$2s = 46$

$\dfrac{2s}{2} = \dfrac{46}{2}$

$s = 23$

The smaller number is 23, so the larger is 23 + 17 = 40.

35. a. base 6, exponent 5

b. base b, exponent 4

37. a. $h^2 h^4 = h^{2+4} = h^6$

b. $\left(m^{10}\right)^2 = m^{20}$

c. $b^2 \cdot b \cdot b^5 = b^{2+1+5} = b^8$

d. $\left(x^3\right)^4 \left(x^2\right)^3 = x^{12} x^6 = x^{18}$

e. $\left(a^2 b^3\right)\left(a^4 b^7\right) = a^{2+4} b^{3+7} = a^6 b^{10}$

f. $\left(12 a^9 b\right)^2 = 12^2 \left(a^9\right)^2 b^2 = 144 a^{18} b^2$

g. $\left(2x^2\right)^3 \left(3x^3\right)^3 = 2^3 \left(x^2\right)^3 3^3 \left(x^3\right)^3$
$= 8x^6 \cdot 27 x^9 = 216 x^{15}$

h. $\left(t^2 t^3\right)^3 = \left(t^5\right)^3 = t^{15}$

Chapter 1 – 8 Cumulative Review

1. a. 7,535,700

b. 7,540,000

3.
$\begin{array}{r} \overset{1}{5},\overset{2}{6}\overset{3}{7}9 \\ 68 \\ 109 \\ +3,458 \\ \hline 9,314 \end{array}$

5.
$$\begin{array}{r} 5{,}345 \\ \times46 \\ \hline 32{,}070 \\ +213{,}800 \\ \hline 245{,}870 \end{array}$$

7. a. $P = 80 + 80 + 50 + 50 = 260\ ft$

   b. $A = 50 \cdot 80 = 4{,}000\ ft^2$

9. a. $1, 2, 4, 5, 10, 20$

   b. $20 = 2 \cdot 10 = 2 \cdot 2 \cdot 5 = 2^2 \cdot 5$

11. $6 + 5\left[20 - \left(3^2 + 1\right)\right]$
    $= 6 + 5\left[20 - (9 + 1)\right]$
    $= 6 + 5\left[20 - 10\right]$
    $= 6 + 5(10)$
    $= 6 + 50$
    $= 56$

13.

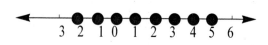

15. a. $-16 + 11 = -5$

    b. $21 - (-17) = 21 + 17 = 38$

    c. $-6(40) = -240$

    d. $\dfrac{-80}{-10} = \dfrac{80}{10} = \dfrac{8}{1} = 8$

17. $\dfrac{(-6)^2 - 1^5}{-4 - 3} = \dfrac{36 - 1}{-7} = \dfrac{35}{-7} = -5$

19. $\dfrac{36}{96} = \dfrac{3 \cdot \cancel{12}}{8 \cdot \cancel{12}} = \dfrac{3}{8}$

21. $\dfrac{10}{21} \cdot \dfrac{3}{10} = \dfrac{\cancel{10} \cdot \cancel{3}}{\cancel{3} \cdot 7 \cdot \cancel{10}} = \dfrac{1}{7}$

23. $\dfrac{1}{9} + \dfrac{5}{6} = \dfrac{1}{9} \cdot \dfrac{2}{2} + \dfrac{5}{6} \cdot \dfrac{3}{3}$
    $= \dfrac{2}{18} + \dfrac{15}{18} = \dfrac{17}{18}$

25. $58\dfrac{4}{11} = 58\dfrac{8}{22} = 57\dfrac{30}{22}$
    $-15\dfrac{1}{2} = -15\dfrac{11}{22} = -15\dfrac{11}{22}$
    $= 42\dfrac{19}{22}$

27. a. $\dfrac{2}{3} + \dfrac{1}{2} \cdot \dfrac{1}{3} = \dfrac{2}{3} + \dfrac{1}{6} = \dfrac{4}{6} + \dfrac{1}{6} = \dfrac{5}{6}$

    b. $1 - \dfrac{1}{6} = \dfrac{6}{6} - \dfrac{5}{6} = \dfrac{1}{6}$ remaining.

29.
$$\begin{array}{r} \overset{1\ \ 1}{6\,4\,5.00000} \\ 9.90005 \\ 0.12000 \\ +\ \ 3.02002 \\ \hline 658.04007 \end{array}$$

31. $-5.8(3.9)(100) = -22.62 \cdot 100 = -2{,}262$

33. Since both terms are negative, the result will be positive. Since there are 3 places left of the decimal in the divisor, move the decimal place 3 units right.
    $-0.4531 \div (-0.001) = 453.1$

35. $270 \div 90 = 27 \div 9 = 3$

37. a. $\dfrac{19}{25} = \dfrac{19}{25} \cdot \dfrac{4}{4} = \dfrac{76}{100} = 0.76$

b.

$$\begin{array}{r} 0.0151 \\ 66\overline{)1.0000} \\ \underline{-66} \\ 340 \\ \underline{-330} \\ 100 \\ \underline{-66} \\ 34 \end{array}$$

$\dfrac{1}{66} = 0.0\overline{15}$

39. $\dfrac{45}{35} = \dfrac{\cancel{5}\cdot 9}{\cancel{5}\cdot 7} = \dfrac{9}{7}$

41. $\dfrac{9.8}{x} = \dfrac{2.8}{5.4}$

$9.8 \cdot 5.4 = 2.8 \cdot x$

$\dfrac{52.92}{2.8} = \dfrac{2.8x}{2.8}$

$18.9 = x$

43. $\dfrac{7{,}500mg}{1} \cdot \dfrac{1g}{1{,}000mg} = \dfrac{7{,}500}{1{,}000}g = 7.5g$

45.

Fraction	Decimal	Percent
$\dfrac{25}{100} = \dfrac{\cancel{25}}{4\cdot\cancel{25}} = \dfrac{1}{4}$	0.25	25%
$\dfrac{1}{3}$	$0.\overline{3}$	$33.\overline{3}\% = 33\dfrac{1}{3}\%$
$\dfrac{21}{500}$	0.042	4.2%

47. $7.8 = 0.12 \cdot x$

$\dfrac{7.8}{0.12} = \dfrac{0.12x}{0.12}$

$65 = x$

49. Use 16,000 as a base. 10% of 16,000 is 1,600 - half that gives 5% which is roughly 800.

51. a. the 18 – 49 age group

b. 41% of 800 = $0.41 \cdot 800 = 328$ people

53. $3x - x^3$

$3 \cdot 4 - 4^3 = 12 - 64 = -52$

55. a. $-3(5x) = (-3 \cdot 5)x = -15x$

b. $-4x(-7x) = (-4(-7))x \cdot x = 28x^2$

57. a. $8x - 3x = 5x$

b. $4a^2 + 6a^2 + 3a^2 - a^2 = 10a^2 + 2a^2 = 12a^2$

c. $4x - 3y - 5x + 2y = -1x - 1y = -x - y$

d. $9(3x - 4) + 2x = 27x - 36 + 2x = 29x - 36$

59. $3x + 2 = -13$

$3x + 2 - 2 = -13 - 2$

$3x = -15$

$\dfrac{3x}{3} = \dfrac{-15}{3}$

$x = -5$

Check:

$3x + 2 = -13$

$3(-5) + 2 \stackrel{?}{=} -13$

$-15 + 2 \stackrel{?}{=} -13$

$-13 = -13$

The solution checks.

61.
$$3(3y-8) = -2(y-4)+3y$$
$$9y-24 = -2y+8+3y$$
$$9y-24 = y+8$$
$$9y-24+24 = y+8+24$$
$$9y = y+32$$
$$9y-y = y+32-y$$
$$8y = 32$$
$$\frac{8y}{8} = \frac{32}{8}$$
$$y = 4$$

Check:

$$3(3y-8) = -2(y-4)+3y$$
$$3(3 \cdot 4-8) \stackrel{?}{=} -2(4-4)+3 \cdot 4$$
$$3(12-8) \stackrel{?}{=} -2 \cdot 0+12$$
$$3 \cdot 4 \stackrel{?}{=} 12$$
$$12 = 12$$

The solution checks.

63. Let x be the number of shifts remaining.

$$37+3x = 100$$
$$37+3x-37 = 100-37$$
$$3x = 63$$
$$\frac{3x}{3} = \frac{63}{3}$$
$$x = 21$$

She must observe for 21 more shifts.

65. a. base 8, exponent 9

b. base a, exponent 3

CHAPTER 9: An Introduction to Geometry

Section 9.1: Basic Geometric Figures; Angles

VOCABULARY

1. Three undefined words in geometry are <u>point</u>, <u>line</u>, and <u>plane</u>.

3. A <u>midpoint</u> divides a line segment into two parts of equal length.

5. An <u>angle</u> is formed by two rays with a common endpoint.

7. A <u>protractor</u> is used to measure angles.

9. The measure of a <u>right</u> angle is $90°$.

11. The measure of a straight angle is $\underline{180°}$.

13. <u>Adjacent</u> angles have the same vertex, are side-by-side, and their interiors do not overlap.

15. When two angles have the same measure, we say that they are <u>congruent</u>.

17. The sum of two complementary angles is $\underline{90°}$.

CONCEPTS

19. a. one

    b. In general, two different points determine exactly one <u>line</u>.

21. a. $\overrightarrow{ST}$, $\overrightarrow{SR}$

    b. S

    c. $\angle RST$, $\angle TSR$, $\angle S$, $\angle 1$

23. a. $\angle$

    b. $\langle$

    c. $\llcorner$

    d. —

25. a.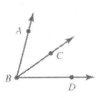

    b.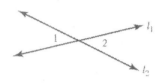

    c.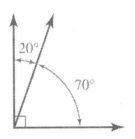

    d.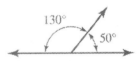

27. The vertical angle property: Vertical angles are <u>congruent</u>.

29. a. false

    b. false

    c. false

    d. true

31. true

33. false

NOTATION

35. The symbol $\overleftrightarrow{AB}$ is read as "<u>line</u> AB."

188 Tussy/Gustafson/Koenig Basic Mathematics for College Students, 4e

37. The symbol $\overrightarrow{AB}$ is read as "ray AB."

39. We read $\angle ABC$ as "angle ABC."

41. The symbol for degree is a small raised circle, °.

43. The symbol $\cong$ is read as "is congruent to."

GUIDED PRACTICE

45. a.

T
•

b.

c.

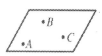

47. a. 2   b. 3

c. 1   d. 6

49. 50°

51. 25°

53. $180° - 105° = 75°$

55. $180° - 50° = 130°$

57. right

59. acute

61. straight

63. obtuse

65. $x = 55° - 45° = 10°$

67. $x = 50° - 22.5° = 27.5°$

69. vertical angles with $\angle OYZ$: 70°

71. vertical angles with $\angle MYX$: 65°

73. $2x = x + 30°$
$2x - x = x + 30° - x$
$x = 30°$

$m(\angle ABD) = 2 \cdot 30° = 60°$
$m(\angle DBE) = 180° - 60° = 120°$

75. $4x + 15° = 7x - 60°$
$4x + 15° + 60° = 7x - 60° + 60°$
$4x + 75° = 7x$
$4x + 75° - 4x = 7x - 4x$
$75° = 3x$
$\dfrac{75°}{3} = \dfrac{3x}{3}$
$25° = x$

$m(\angle ZYQ) = 7 \cdot 25° - 60° = 175° - 60° = 115°$
$m(\angle PYQ) = 180° - 115° = 65°$

77. $x + 30° = 90°$
$x + 30° - 30° = 90° - 30°$
$x = 60°$

79. $x + 105° = 180°$
$x + 105° - 105° = 180° - 105°$
$x = 75°$

TRY IT YOURSELF

81. a. true

b. False, it is a segment so it has two endpoints.

c. False, it is a line so it has no endpoints.

d. False, G is the vertex of that angle.

e. true

f. true

83. 40°

85. 135°

87. a. vertical angles:

$m(\angle 3) = 50°$

b. supplemental angles:

$m(\angle 1) + m(\angle 4) = 180°$
$50° + m(\angle 4) = 180°$
$50° + m(\angle 4) - 50° = 180° - 50°$
$m(\angle 4) = 130°$

c.

$m(\angle 2) = m(\angle 4) = 130°$
$m(\angle 1) + m(\angle 2) + m(\angle 3)$
$= 50° + 130° + 50° = 230°$

d.

$m(\angle 2) + m(\angle 4) = 130° + 130° = 260°$

89. a.

$m(\angle BAC) = 24°$
$90° - 24° = 66°$

b.

$m(\angle BAC) = 24°$
$180° - 24° = 156°$

91. complement: $90° - 51° = 39°$

supplement of that: $180° - 39° = 141°$

93. complement: $90° - 1° = 89°$

complement of that: $90° - 89° = 1°$

APPLICATIONS

95. a. about $80°$

b. about $30°$

c. about $65°$

97. a. $90 - 63 = 27°$ from horizontal

b. $180 - 150 = 30°$ with the ground

WRITING

99. a. It means he completely reversed direction and went directly the other way.

b. She flipped completely over and landed facing in the same direction as she started.

101. Yes, if both angles are $45°$ they are both complementary and equal.

REVIEW

103. $\dfrac{1}{2} + \dfrac{2}{3} + \dfrac{3}{4} = \dfrac{1}{2} \cdot \dfrac{6}{6} + \dfrac{2}{3} \cdot \dfrac{4}{4} + \dfrac{3}{4} \cdot \dfrac{3}{3}$

$= \dfrac{6}{12} + \dfrac{8}{12} + \dfrac{9}{12} = \dfrac{23}{12} = 1\dfrac{11}{12}$

105. $\dfrac{5}{8} \cdot \dfrac{2}{15} \cdot \dfrac{6}{5} = \dfrac{\cancel{5} \cdot \cancel{2} \cdot \cancel{2} \cdot \cancel{3}}{\cancel{2} \cdot \cancel{2} \cdot 2 \cdot \cancel{3} \cdot \cancel{5} \cdot 5}$

$= \dfrac{1}{2 \cdot 5} = \dfrac{1}{10}$

Section 9.2: Parallel and Perpendicular Lines

VOCABULARY

1. Two lines that lie in the same plane are called coplanar. Two lines that lie in different planes are called noncoplanar.

3. Perpendicular lines are lines that intersect and form right angles.

5. $\angle 4$ and $\angle 6$ are alternate interior angles.

CONCEPTS

7. a.

b.

9. a.

b.

11. When two parallel lines are cut by a transversal, <u>corresponding</u> angles are congruent.

13. When two parallel lines are cut by a transversal, <u>interior</u> angles on the same side of the transversal are supplementary.

15. They are perpendicular.

NOTATION

17. The symbol ⌐ indicates a <u>right</u> angle.

19. The symbol ⊥ is read "is <u>perpendicular</u> to."

GUIDED PRACTICE

21. a. ∠1 & ∠5 ; ∠4 & ∠8 ;
       ∠2 & ∠6 ; ∠3 & ∠7

    b. ∠3 , ∠4 , ∠5 , ∠6

    c. ∠3 & ∠5 ; ∠4 & ∠6

23. By corresponding angles and vertical angles:

    $m(\angle 1) = m(\angle 4) = m(\angle 5) = m(\angle 8) = 130°$

    By supplemental angles, since $180° - 130° = 50°$,

    $m(\angle 2) = m(\angle 3) = m(\angle 6) = m(\angle 7) = 50°$

25. ∠1 ≅ ∠X ; ∠2 ≅ ∠N

27. $4x - 8° = 2x + 16°$
    $4x - 8° + 8° = 2x + 16° + 8°$
    $4x = 2x + 24°$
    $4x - 2x = 2x + 24° - 2x$
    $2x = 24°$
    $\dfrac{2x}{2} = \dfrac{24°}{2}$
    $x = 12°$

    $4x - 8 = 4(12) - 8 = 48 - 8 = 40°$

    $2x + 16 = 2(12) + 16 = 24 + 16 = 40°$

29. $5x + 6x + 70° = 180°$
    $11x + 70° = 180°$
    $11x + 70° - 70° = 180° - 70°$
    $11x = 110°$
    $\dfrac{11x}{11} = \dfrac{110°}{11}$
    $x = 10°$

    $5(10°) = 50°$

    $6(10°) + 70° = 60° + 70° = 130°$

TRY IT YOURSELF

31. a. alternate interior: $m(\angle 1) = 50°$

    alternate interior: $m(\angle 3) = 45°$

    supplement: $m(\angle 2) + 45° = 180°$
    $m(\angle 2) = 135°$

$$50° + m(\angle 4) + m(\angle 3) = 180°$$

lastly:

$$50° + m(\angle 4) + 45° = 180°$$
$$m(\angle 4) + 95° = 180°$$
$$m(\angle 4) = 85°$$

b. They form a straight angle, so $180°$.

c. They form a triangle, so $180°$.

33. vertical angles: $\angle 1 \cong \angle 2$

alternate interior: $\angle A \cong \angle E$; $\angle B \cong \angle D$

35.
$$(3x + 20°) + (x) = 180°$$
$$4x + 20° = 180°$$
$$4x + 20° - 20° = 180° - 20°$$
$$4x = 160°$$
$$\frac{4x}{4} = \frac{160°}{4}$$
$$x = 40°$$
$$3x + 20° = 3(40°) + 20° = 120° + 20° = 140°$$

37.
$$6x - 2° = 9x - 38°$$
$$6x - 2° + 38° = 9x - 38° + 38°$$
$$6x + 36° = 9x$$
$$6x + 36° + 6x = 9x - 6x$$
$$36° = 3x$$
$$\frac{36°}{3} = \frac{3x}{3}$$
$$12° = x$$
$$6x - 2° = 6(12°) - 2° = 72° - 2° = 70°$$
$$9x - 38° = 9(12°) - 38° = 108° - 38° = 70°$$

APPLICATIONS

39. The plummet string should be perpendicular to the top of the stones.

41.
$$\angle DCF \cong \angle ABC$$
$$m(\angle ABE) + m(\angle ABC) = 180°$$
$$m(\angle ABE) + 130° = 180°$$
$$m(\angle ABE) = 150°$$

43. Wallpaper strips need to be perpendicular to the ceiling (and floor), and must also be parallel to one another to preserve the pattern.

45. $m(\angle 2) = 105°$ (corresponding angles)

$$105° + m(\angle 1) = 180°$$
$$m(\angle 1) = 75°$$

$m(\angle 3) = 75°$ (corresponding angles with $\angle 1$)

WRITING

47. On the North side of the street cars will park parallel to the West – East line. On the South side of the street East of the planter cars will park perpendicular to the West – East line. On the South side of the street West of the planter the cars will park at slant to the West – East line.

49. Since they are corresponding angles, $\angle CBE \cong \angle FEH$.

Since they are vertical angles, $\angle CBE \cong \angle ABD$.

Therefore, they all have the same measurement of $100°$

51. No – only if the transversal crosses parallel lines.

REVIEW

53.
$$\frac{60}{100} = \frac{x}{120}$$
$$60 \cdot 120 = x \cdot 100$$
$$\frac{7200}{100} = \frac{x \cdot 100}{100}$$
$$72 = x$$

55.
$$\frac{x}{100} = \frac{225}{500}$$
$$500 \cdot x = 225 \cdot 100$$
$$\frac{500 \cdot x}{500} = \frac{22,500}{500}$$
$$x = 45$$

225 is 45% of 500.

57. yes

© 2011 Cengage Learning. All Rights Reserved. May not be scanned, copied or duplicated, or posted to a publicly accessible website, in whole or in part.

59. $\dfrac{4oz}{12oz} = \dfrac{\cancel{4\,oz}}{3\cdot \cancel{4\,oz}} = \dfrac{1}{3}$

Section 9.3: Triangles

VOCABULARY

1. A <u>polygon</u> is a closed geometric figure with at least three line segments for its sides.

3. A point where two sides of a polygon intersect is called a <u>vertex</u> of the polygon.

5. A triangles with three sides of equal length is called an <u>equilateral</u> triangle. An <u>isosceles</u> triangle has at least two sides of equal length. A <u>scalene</u> triangle has no sides of equal length.

7. The longest side of a right triangle is called the <u>hypotenuse</u>. The other two sides of a right triangle are called <u>legs</u>.

9. In this section we discussed the sum of the measures of the angles of a triangle. The word *sum* indicates the operation of <u>addition</u>.

CONCEPTS

11. a. b. c.

 d. e. f.

13. a. b. c.

 d. e. f.

15. a. 90°

 b. right

 c. $\overline{AB}$, $\overline{BC}$

 d. $\overline{AC}$

 e. $\overline{AC}$

 f. $\overline{AC}$

17. a. The <u>isosceles</u> triangle theorem states that if two sides of a triangle are congruent, then the angles opposite those sides are congruent.

 b. The <u>converse</u> of the isosceles triangle theorem states that if two angles of a triangle are congruent, then the sides opposite the angles have the same length, and the triangle is isosceles.

19. a.

 b. isosceles

NOTATION

21. The symbol △ means <u>triangle</u>.

23.

GUIDED PRACTICE

25. a. 4, quadrilateral, 4

 b. 6, hexagon, 6

27. a. 7, heptagon, 7

 b. 9, nonagon, 9

29. a. scalene

 b. isosceles

31. a. equilateral

 b. scalene

33. Yes, it has two equal angles.

35. no

37. $35° + 90° + y = 180°$
$125° + y = 180°$
$125° + y - 125° = 180° - 125°$
$y = 55°$

39. $45° + 90° + y = 180°$
$135° + y = 180°$
$135° + y - 135° = 180° - 135°$
$y = 45°$

41. $(x + 10°) + (x + 20°) + x = 180°$
$3x + 30° = 180°$
$3x + 30° - 30° = 180° - 30°$
$3x = 150°$
$\dfrac{3x}{3} = \dfrac{150°}{3}$
$x = 50°$
$50° + 10° = 60°$
$50° + 20° = 70°$

43. $4x + 4x + x = 180°$
$9x = 180°$
$\dfrac{9x}{9} = \dfrac{180°}{9}$
$x = 20°$
$4x = 4(20°) = 80°$

45. $56° + 56° + x = 180°$
$112° + x = 180°$
$112° + x - 112° = 180° - 112°$
$x = 68°$

47. $85.5° + 85.5° + x = 180°$
$171° + x = 180°$
$171° + x - 171° = 180° - 171°$
$x = 9°$

49. $x + x + 102° = 180°$
$2x + 102° = 180°$
$2x + 102° - 102° = 180° - 102°$
$2x = 78°$
$\dfrac{2x}{2} = \dfrac{78°}{2}$
$x = 39°$

51. $x + x + 90.5° = 180°$
$2x + 90.5° = 180°$
$2x + 90.5° - 90.5° = 180° - 90.5°$
$2x = 89.5°$
$\dfrac{2x}{2} = \dfrac{89.5°}{2}$
$x = 44.75°$

TRY IT YOURSELF

53. $x + 76° + 76° = 180°$
$x + 152° = 180°$
$x + 152° - 152° = 180° - 152°$
$x = 28°$

55. $x + 53.5° + 53.5° = 180°$
$x + 107° = 180°$
$x + 107° - 107° = 180° - 107°$
$x = 73°$

57. $30° + 60° + x = 180°$
$90° + x = 180°$
$90° + x - 90° = 180° - 90°$
$x = 90°$

59. $100° + 35° + x = 180°$
$135° + x = 180°$
$135° + x - 135° = 180° - 135°$
$x = 45°$

61. $25.5° + 63.8° + x = 180°$
$89.3° + x = 180°$
$89.3° + x - 89.3° = 180° - 89.3°$
$x = 90.7°$

63.
$$29° + 89.5° + x = 180°$$
$$118.5° + x = 180°$$
$$118.5° + x - 118.5° = 180° - 118.5°$$
$$x = 61.5°$$

65.
$$x + x + 156° = 180°$$
$$2x + 156° = 180°$$
$$2x + 156° - 156° = 180° - 156°$$
$$2x = 24°$$
$$\frac{2x}{2} = \frac{24°}{2}$$
$$x = 12°$$

67.
$$x + x + 75° = 180°$$
$$2x + 75° = 180°$$
$$2x + 75° - 75° = 180° - 75°$$
$$2x = 105°$$
$$\frac{2x}{2} = \frac{105°}{2}$$
$$x = 52.5°$$

69. If it is a base angle:
$$39° + 39° + x = 180°$$
$$78° + x = 180°$$
$$x = 102°$$
$$39°, 39°, 102°$$

If it is the vertex angle:
$$x + x + 39° = 180°$$
$$2x + 39° = 180°$$
$$2x = 141°$$
$$x = 70.5°$$
$$39°, 70.5°, 70.5°$$

71.
$$m(\angle DBA) = 180° - 73° - 61° = 46°$$
$$46° + 49° + m(\angle EBC) = 180°$$
$$95° + m(\angle EBC) = 180°$$
$$m(\angle EBC) = 85°$$
$$85° + 22° + m(\angle C) = 180°$$
$$107° + m(\angle C) = 180°$$
$$m(\angle C) = 73°$$

73.
$$m(\angle ONM) = 180° - 79° = 101°$$
$$m(\angle NOM) = 180° - 64° - 101° = 15°$$
$$m(\angle NOM) + m(\angle NOQ) + 90° = 180°$$
$$15° + m(\angle NOQ) + 90° = 180°$$
$$105° + m(\angle NOQ) = 180°$$
$$m(\angle NOQ) = 75°$$

APPLICATIONS

75. a. octagon

b. triangle

c. pentagon

77. As the jack is raised, the two sides remain the same length. (Otherwise the car could tip!)

79. equilateral

WRITING

81. The Pentagon is the headquarters of the U.S. Military. It is an enormous building with 5 sides, hence the name.

83. A triangle can not have two right angles because that would equal the full $180°$ and not leave anything for the third required angle.

REVIEW

85.
$$\frac{20}{100} = \frac{x}{110}$$
$$20 \cdot 110 = x \cdot 100$$
$$\frac{2,200}{100} = \frac{x \cdot 100}{100}$$
$$22 = x$$

87. $\dfrac{x}{100} = \dfrac{80}{200}$

$x \cdot 200 = 80 \cdot 100$

$\dfrac{x \cdot 200}{200} = \dfrac{8{,}000}{200}$

$x = 40$

800 is 40% of 200.

89. $0.85 \div 2(0.25)$

$= 0.425(0.25)$

$= 0.10625$

Section 9.4: The Pythagorean Theorem

VOCABULARY

1. In a right triangle, the side opposite the $90°$ angle is called the <u>hypotenuse</u>. The other two sides are called the <u>legs</u>.

3. The <u>Pythagorean</u> theorem states that in any right triangle, the square of the length of the hypotenuse is equal to the sum of the squares of the lengths of the two legs.

CONCEPTS

5. If a and b are the lengths of two legs of a right triangle and c is the hypotenuse, then $a^2 + b^2 = c^2$.

7. The converse of the Pythagorean theorem: If a triangle has three sides of lengths $a, b,$ and c, such that $a^2 + b^2 = c^2$, then the triangle is a right triangle.

9. a. $\overline{BC}$

   b. $\overline{AB}$

   c. $\overline{AC}$

NOTATION

11. $8^2 + 6^2 = c^2$

$64 + 36 = c^2$

$100 = c^2$

$\sqrt{100} = c$

$10 = c$

GUIDED PRACTICE

13. $6^2 + 8^2 = c^2$

$36 + 64 = c^2$

$100 = c^2$

$\sqrt{100} = c$

$10\,ft = c$

15. $5^2 + 12^2 = c^2$

$25 + 144 = c^2$

$169 = c^2$

$\sqrt{169} = c$

$13m = c$

17. $48^2 + 55^2 = c^2$

$2{,}304 + 3{,}025 = c^2$

$5{,}329 = c^2$

$\sqrt{5{,}329} = c$

$73mi = c$

19. $88^2 + 105^2 = c^2$

$7{,}744 + 11{,}025 = c^2$

$18{,}769 = c^2$

$\sqrt{18{,}769} = c$

$137cm = c$

21. $10^2 + b^2 = 26^2$

$100 + b^2 = 676$

$100 + b^2 - 100 = 676 - 100$

$b^2 = 576$

$b = \sqrt{576}$

$b = 24cm$

23. $a^2 + 18^2 = 82^2$
$a^2 + 324 = 6,724$
$a^2 + 324 - 324 = 6,724 - 324$
$a^2 = 6,400$
$a = \sqrt{6,400}$
$a = 80m$

25. $a^2 + 21^2 = 29^2$
$a^2 + 441 = 841$
$a^2 + 441 - 441 = 841 - 441$
$a^2 = 400$
$a = \sqrt{400}$
$a = 20m$

27. $180^2 + b^2 = 181^2$
$32,400 + b^2 = 32,761$
$32,400 + b^2 - 32,400 = 32,761 - 32,400$
$b^2 = 361$
$b = 19m$

29. $5^2 + b^2 = 6^2$
$25 + b^2 = 36$
$25 + b^2 - 25 = 36 - 25$
$b^2 = 11$
$b = \sqrt{11} cm \approx 3.32 cm$

31. $12^2 + 8^2 = c^2$
$144 + 64 = c^2$
$208 = c^2$
$\sqrt{208} m = c \approx 14.42 m$

33. $9^2 + 3^2 = c^2$
$81 + 9 = c^2$
$90 = c^2$
$\sqrt{90} in = c \approx 9.49 in$

35. $a^2 + 4^2 = 6^2$
$a^2 + 16 = 36$
$a^2 + 16 - 16 = 36 - 16$
$a^2 = 20$
$a = \sqrt{20} in \approx 4.47 in$

37. $12^2 + 14^2 \stackrel{?}{=} 15^2$
$144 + 196 \stackrel{?}{=} 225$
$400 = 225$

Not a right triangle.

39. $33^2 + 56^2 \stackrel{?}{=} 65^2$
$1,089 + 3,136 \stackrel{?}{=} 4,225$
$4,225 = 4,225$

It is a right triangle.

APPLICATIONS

41. $20^2 = 16^2 + x^2$
$400 = 256 + x^2$
$400 - 256 = 256 + x^2 - 256$
$144 = x^2$
$\sqrt{144} = x$
$12 ft = x$

43. $20^2 + 15^2 = x^2$
$400 + 225 = x^2$
$625 = x^2$
$\sqrt{625} = x$
$25 in = x$

45.
$$90^2 + 90^2 = x^2$$
$$8,100 + 8,100 = x^2$$
$$16,200 = x^2$$
$$\sqrt{16,200} = x$$
$$127.28\,ft \approx x$$

47.
$$9^2 + h^2 = 37^2$$
$$81 + h^2 = 1,369$$
$$81 + h^2 - 81 = 1,369 - 81$$
$$h^2 = 1,288$$
$$h = \sqrt{1,288} \approx 35.9\,ft$$

The ladder will reach the window.

WRITING

49. If you add the squares of the shorter sides of a right triangle, you get the square of the longest side.

51. If you count the total number of little squares in the two smaller squares you get the same number as the number of little squares in the large square.

REVIEW

53.
$$2b + 3 = -15$$
$$2(-8) + 3 \stackrel{?}{=} -15$$
$$-16 + 3 \stackrel{?}{=} -15$$
$$-13 = -15$$

-8 is not a solution.

55.
$$0.5x = 2.9$$
$$0.5(5) \stackrel{?}{=} 2.9$$
$$2.5 = 2.9$$

5 is not a solution.

57.
$$33 - \frac{x}{2} = 30$$
$$33 - \frac{(-6)}{2} \stackrel{?}{=} 30$$
$$33 - (-3) \stackrel{?}{=} 30$$
$$36 = 30$$

-6 is not a solution.

59.
$$3x - 2 = 4x - 5$$
$$3(12) - 2 \stackrel{?}{=} 4(12) - 5$$
$$36 - 2 \stackrel{?}{=} 48 - 5$$
$$33 = 43$$

12 is not a solution.

Section 9.5: Congruent Triangles and Similar Triangles

VOCABULARY

1. <u>Congruent</u> triangles are the same size and the same shape.

3. Two angles or two line segments with the same measure are said to be <u>congruent</u>.

5. If two triangles are <u>similar</u>, they have the same shape but not necessarily the same size.

CONCEPTS

7. a. No, they are different sizes.

b. Yes, they have the same shape.

9. $\triangle XYZ \cong \triangle PRQ$

11. $\triangle RST \sim \triangle MNO$

13. $\angle A \cong \angle B$, $\angle Y \cong \angle T$, $\angle Z \cong \angle R$
$\overline{YZ} \cong \overline{TR}$, $\overline{AZ} \cong \overline{BR}$, $\overline{AY} \cong \overline{BT}$

15. Two triangles are <u>congruent</u> if and only if their vertices can be matched so that the corresponding sides and the corresponding angles are congruent.

17. SAS property: If two sides and the <u>angle</u> between them in one triangle are congruent, respectively, to two sides and the <u>angle</u> between them in a second triangle, the triangles are congruent.

19.
$$\frac{x}{15} = \frac{20}{3}$$
$$x \cdot 3 = 20 \cdot 15$$
$$\frac{x \cdot 3}{3} = \frac{300}{3}$$
$$x = 100$$

21. $\dfrac{h}{2.6} = \dfrac{27}{13}$

$h \cdot 13 = 27 \cdot 2.6$

$\dfrac{h \cdot 13}{13} = \dfrac{70.2}{13}$

$h = 5.4$

23. Two triangles are similar if and only if their vertices can be matched so that corresponding angles are congruent and the lengths of corresponding sides are <u>proportional</u>.

25. Congruent triangles are always similar, but similar triangles are not always <u>congruent</u>.

NOTATION

27. The symbol $\cong$ is read as "is congruent to."

29.

GUIDED PRACTICE

31. $\overline{AC} \cong \overline{DF}$

$\overline{DE} \cong \overline{AB}$

$\overline{BC} \cong \overline{EF}$

$\angle A \cong \angle D$

$\angle E \cong \angle B$

$\angle F \cong \angle C$

33. a.

$\overline{BC} \cong \overline{MN}$

$\overline{CD} \cong \overline{NO}$

$\overline{BD} \cong \overline{MO}$

$\angle B \cong \angle M$

$\angle C \cong \angle N$

$\angle D \cong \angle O$

b. 72°

c. 10 ft

d. 9 ft

35. yes: SSS

37. Not necessarily. Despite appearances the other sides for SAS (or angle for ASA) may not be of equal size.

39. a.

$\angle L \cong \angle H$

$\angle M \cong \angle J$

$\angle R \cong \angle E$

b.

$\dfrac{LM}{HJ} = \dfrac{MR}{JE}, \ \dfrac{MR}{JE} = \dfrac{LR}{HE}, \ \dfrac{LM}{HJ} = \dfrac{LR}{HE}$

c.

$\dfrac{LM}{HJ} = \dfrac{MR}{JE} = \dfrac{LR}{HE}$

41. yes: vertical angle, right angle, third angle

43. Not necessarily – two angles are needed.

45. yes

47. Not necessarily – they may not be right triangles.

49. yes

51. not necessarily

53. $\dfrac{x}{28} = \dfrac{6}{21}$

$x \cdot 21 = 6 \cdot 28$

$\dfrac{x \cdot 21}{21} = \dfrac{168}{21}$

$x = 8$

$\dfrac{y}{10} = \dfrac{21}{6}$

$y \cdot 6 = 21 \cdot 10$

$\dfrac{y \cdot 6}{6} = \dfrac{210}{6}$

$y = 35$

55.
$$\frac{x}{40} = \frac{75}{50}$$
$$x \cdot 50 = 75 \cdot 40$$
$$\frac{x \cdot 50}{50} = \frac{3{,}000}{50}$$
$$x = 60$$

$$\frac{y}{57} = \frac{50}{75}$$
$$y \cdot 75 = 50 \cdot 57$$
$$\frac{y \cdot 75}{75} = \frac{2{,}850}{75}$$
$$y = 38$$

TRY IT YOURSELF

57. true: SSS

59. False: the angle needs to be the included angle.

61. yes: SSS

63. yes: SAS

65. yes: ASA (alternate interior angles)

67. not necessarily

69. $x = 80°$
$y = 2\,yd$

71. $x = 19°$
$y = 14\,m$

73. $x = 6\,mm$

75. $x = 50°$

77.
$$\frac{x}{10} = \frac{5}{12}$$
$$x \cdot 12 = 5 \cdot 10$$
$$\frac{x \cdot 12}{12} = \frac{50}{12}$$
$$x = \frac{25}{6} = 4\frac{1}{6}$$

79.
$$\frac{x}{12} = \frac{x+4}{15}$$
$$x \cdot 15 = (x+4)12$$
$$15x = 12x + 48$$
$$15x - 12x = 12x + 48 - 12x$$
$$3x = 48$$
$$\frac{3x}{3} = \frac{48}{3}$$
$$x = 16$$

APPLICATIONS

81. $m(\overline{AO}) = m(\overline{CO}) = 9.5\,cm$
$9.5 + 8 = 17.5\,cm$

83.
$$\frac{25}{74} = \frac{20}{w}$$
$$25 \cdot w = 20 \cdot 74$$
$$\frac{25 \cdot w}{25} = \frac{1480}{25}$$
$$w = 59.2\,ft$$

85.
$$\frac{6}{4} = \frac{t}{24}$$
$$6 \cdot 24 = t \cdot 4$$
$$\frac{144}{4} = \frac{t \cdot 4}{4}$$
$$36\,ft = t$$

87.
$$\frac{3}{2.5} = \frac{h}{29}$$
$$3 \cdot 29 = h \cdot 2.5$$
$$\frac{87}{2.5} = \frac{h \cdot 2.5}{2.5}$$
$$34.8\,ft = h$$

89.
$$\frac{200}{1,000} = \frac{x}{5,280}$$
$$\frac{1}{5} = \frac{x}{5,280}$$
$$1 \cdot 5,280 = x \cdot 5$$
$$\frac{5,280}{5} = x$$
$$1,056 \, ft = x$$

WRITING

91. There is no SSA property for triangles because the angle could be opening in one direction in the first triangle and in the other direction on the second triangle – the included angle would be different in each case.

REVIEW

93.
$$21 = 3 \cdot 7$$
$$27 = 3 \cdot 3 \cdot 3$$
$$LCM(21, 27) = 3 \cdot 3 \cdot 3 \cdot 7 = 189$$

95.
$$63 = 3 \cdot 3 \cdot 7$$
$$84 = 2 \cdot 2 \cdot 3 \cdot 7$$
$$GCF(63, 84) = 3 \cdot 7 = 21$$

Section 9.6: Quadrilaterals and Other Polygons

VOCABULARY

1. A <u>quadrilateral</u> is a polygon with four sides.

3. A <u>rectangle</u> is a quadrilateral with four right angles.

5. A <u>rhombus</u> is a parallelogram with four sides of equal length.

7. A <u>trapezoid</u> has two sides that are parallel and two sides that are not parallel. The parallel sides are called <u>bases</u>. The legs of an <u>isosceles</u> trapezoid have the same length.

CONCEPTS

9. a. four; A, B, C, D

   b. four; $\overline{AB}, \overline{BC}, \overline{CD}, \overline{DA}$

   c. two; $\overline{AC}, \overline{BD}$

   d. yes, no, no, yes (The sides must be consecutive.)

11. a. $\overline{ST} \parallel \overline{VU}$ b. $\overline{SV} \parallel \overline{TU}$

13. a. All four angles are <u>right</u> angles.

    b. Opposite sides are <u>parallel</u>.

    c. Opposite sides have equal <u>length</u>.

    d. The diagonals have equal <u>length</u>.

    e. The diagonals intersect at their <u>midpoint</u>.

15. It is a rectangle.

17. a. no

    b. yes

    c. no

    d. yes

    e. no

    f. yes

19. a. isosceles

    b. $\angle J, \angle M$

    c. $\angle K, \angle L$

    d. $m(\angle J) = m(\angle M)$
       $m(\angle K) = m(\angle L)$
       $m(\overline{JK}) = m(\overline{ML})$

NOTATION

21. The four sides of the quadrilateral have the same length.

23. S represents the sum of the angles in the polygon.
    n represents the number of sides.

GUIDED PRACTICE

25. a. square

    b. rhombus

    c. trapezoid

    d. rectangle

27. a. 90°

    b. 9 (diagonals cross at the midpoint)

    c. $9 + 9 = 18$

    d. 18 (diagonals are the same length)

29. Interior angles are supplemental:

    $138° + x = 180°$

    a. $138° + x - 138° = 180° - 138°$
    $x = 42°$

    $y + 85° = 180°$

    b. $y + 85° - 85° = 180° - 85°$
    $y = 95°$

31. a. 9

    b. 70°

    c. $m(\angle x) + m(\angle y) = 180°$
    $70° + m(\angle y) = 180°$
    $70° + m(\angle y) - 70° = 180° - 70°$
    $m(\angle y) = 110°$

    d. 110°

33. $S = (n-2)180° = (14-2)180°$
    $= (12)180° = 2,160°$

35. $S = (n-2)180° = (20-2)180°$
    $= (18)180° = 3,240°$

37. $S = (n-2)180° = (8-2)180°$
    $= (6)180° = 1,080°$

39. $S = (n-2)180° = (12-2)180°$
    $= (10)180° = 1,800°$

41. $540° = (n-2)180°$
    $540° = 180°n - 360°$
    $540° + 360° = 180°n - 360° + 360°$
    $900° = 180°n$
    $\dfrac{900°}{180°} = \dfrac{180°n}{180°}$
    $5 = n$

43. $900° = (n-2)180°$
    $900° = 180°n - 360°$
    $900° + 360° = 180°n - 360° + 360°$
    $1,260° = 180°n$
    $\dfrac{1,260°}{180°} = \dfrac{180°n}{180°}$
    $7 = n$

45. $1,980° = (n-2)180°$
    $1,980° = 180°n - 360°$
    $1,980° + 360° = 180°n - 360° + 360°$
    $2,340° = 180°n$
    $\dfrac{2,340°}{180°} = \dfrac{180°n}{180°}$
    $13 = n$

47. $2,160° = (n-2)180°$
    $2,160° = 180°n - 360°$
    $2,160° + 360° = 180°n - 360° + 360°$
    $2,520° = 180°n$
    $\dfrac{2,520°}{180°} = \dfrac{180°n}{180°}$
    $14 = n$

TRY IT YOURSELF

49. $m(\angle 1) + 60° = 90°$

    a. $m(\angle 1) + 60° - 60° = 90° - 60°$
    $m(\angle 1) = 30°$

    b. alternate interior: $m(\angle 3) = m(\angle 1) = 30°$

c. alternate interior: $m(\angle 2) = 60°$

d. diagonals are of equal length: 8cm

e. $\dfrac{1}{2}(8) = 4cm$

51. $S = (n-2)180° = (4-2)180° = (2)180° = 360°$

$2x + (2x+10°) + (3x+30)° + x = 360°$
$8x + 40° = 360°$
$8x + 40° - 40° = 360° - 40°$
$8x = 320°$
$\dfrac{8x}{8} = \dfrac{320°}{8}$
$x = 40°$

$x = 40°$
$2x = 2 \cdot 40° = 80°$
$2x + 10° = 2 \cdot 40° + 10° = 90°$
$3x + 30° = 3 \cdot 40° + 30° = 150°$

APPLICATIONS

53. a. trapezoid

b. square

c. rectangle

d. trapezoid

e. parallelogram

55. $S = (n-2)180° = (5-2)180°$
$= 3(180°) = 540°$

WRITING

57. A square is a rectangle because its opposite sides are parallel and all the angles are right angles.

59. She can check that the diagonals are of the same length.

REVIEW

61. two hundred fifty-four thousand, three hundred nine

63. eighty-two million, four hundred fifteen

Section 9.7: Perimeters and Areas of Polygons

VOCABULARY

1. The distance around a polygon is called the perimeter.

3. The measure of the surface enclosed by a polygon is called its area.

5. The area of a polygon is measured in square units.

CONCEPTS

7. $8\,ft \cdot 16\,ft = 128\,ft^2$

9. a. $P = 4s$ b. $P = 2l + 2w$

11.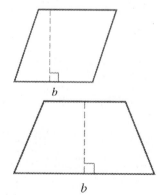

13. a rectangle and a triangle

NOTATION

15. a. The symbol $1in^2$ means one <u>square</u> inch.

b. One square meter is expressed as $1m^2$.

GUIDED PRACTICE

17. $P = 4s = 4 \cdot 8 = 32 in$

19. $P = 4s = 4 \cdot 5.75 = 23 mi$

21. First convert: $2 ft = 24 in$

$P = 2l + 2w = 2 \cdot 24 + 2 \cdot 7$
$= 48 + 14 = 62 in$

23. First convert: $3 ft = 36 in$

$P = 2l + 2w = 2 \cdot 11 + 2 \cdot 36$
$= 22 + 72 = 94 in$

25. Let x = the length of the third side.

$10 + 10 + x = 35$
$20 + x = 35$
$20 + x - 20 = 35 - 20$
$x = 15 ft$

27. Let x = the length of the legs.

$x + 10 + x + 15 = 35$
$2x + 25 = 35$
$2x + 25 - 25 = 35 - 25$
$2x = 10$
$\dfrac{2x}{2} = \dfrac{10}{2}$
$x = 5m$

29. $A = s^2 = 4^2 = 16 cm^2$

31. $A = s^2 = 2.5^2 = 6.25 m^2$

33. $1 ft^2 = 1 ft \cdot 1 ft = 12 in \cdot 12 in = 144 in^2$

35. $1 m^2 = 1m \cdot 1m = 1,000 mm \cdot 1,000 mm$
$= 1,000,000 mm^2$

37. $1 mi^2 = 1mi \cdot 1mi = 5,280 ft \cdot 5,280 ft$
$= 27,878,400 ft^2$

39. $1 km^2 = 1km \cdot 1km = 1,000m \cdot 1,000m$
$= 1,000,000 m^2$

41. First convert to feet: the dimensions are 9ft by 15ft.

$A = lw = 9 \cdot 15 = 135 ft^2$

43. First convert to feet: the dimensions are 60ft by 186ft.

$A = lw = 60 \cdot 186 = 11,160 ft^2$

45. $A = \dfrac{1}{2} bh = \dfrac{1}{2} \cdot 10 \cdot 5 = 25 in^2$

47. $A = \dfrac{1}{2} bh = \dfrac{1}{2} \cdot 9 \cdot 6 = 27 cm^2$

49. $A = \dfrac{1}{2} bh = \dfrac{1}{2} \cdot 5 \cdot 3 = 7.5 in^2$

51. $A = \dfrac{1}{2} bh = \dfrac{1}{2} \cdot 7 \cdot 3 = 10.5 mi^2$

53. $A = \dfrac{1}{2} h(b_1 + b_2) = \dfrac{1}{2} \cdot 4(8 + 12)$
$= 2(20) = 40 ft^2$

55. $A = \dfrac{1}{2} h(b_1 + b_2) = \dfrac{1}{2} \cdot 7(10 + 16)$
$= \dfrac{7}{2}(26) = 91 cm^2$

57. $A = bh$
$60 = b \cdot 15$
$\dfrac{60}{15} = \dfrac{b \cdot 15}{15}$
$4m = b$

59.
$$A = lw$$
$$36 = 3 \cdot w$$
$$\frac{36}{3} = \frac{3 \cdot w}{3}$$
$$12cm = w$$

61.
$$A = \frac{1}{2}bh$$
$$54 = \frac{1}{2} \cdot 3 \cdot h$$
$$54 = \frac{3}{2} \cdot h$$
$$\frac{2}{3} \cdot 54 = \frac{2}{3} \cdot \frac{3}{2} \cdot h$$
$$36m = h$$

63.
$$P = 2l + 2w$$
$$64 = 2 \cdot 21 + 2w$$
$$64 = 42 + 2w$$
$$64 - 42 = 42 + 2w - 42$$
$$22 = 2w$$
$$\frac{22}{2} = \frac{2w}{2}$$
$$11mi = w$$

65. Break down the figure into a rectangle of dimensions 12in by 6in and a triangle with base 12in and height 5in.
$$A = \frac{1}{2}bh + lw = \frac{1}{2} \cdot 12 \cdot 5 + 12 \cdot 6$$
$$= 30 + 72 = 102 in^2$$

67.
$$A = \frac{1}{2}bh + lw = \frac{1}{2} \cdot 30 \cdot 20 + 30 \cdot 2$$
$$= 300 + 60 = 360 ft^2$$

69. Outside rectangle: $A = lw = 6 \cdot 14 = 84 m^2$

Inside square: $A = lw = 3 \cdot 3 = 9 m^2$

Difference: $84 - 9 = 75 m^2$

71. Outside square: $A = s^2 = 10^2 = 100 yd^2$

Inside triangle: $A = \frac{1}{2}bh = \frac{1}{2} \cdot 10 \cdot 5 = 25 yd^2$

Difference: $100 - 25 = 75 yd^2$

73. Area of the floor: $A = lw = 8 \cdot 5 = 40 yd^2$

$40 \cdot \$30 = \$1,200$ for all the flooring.

75. Perimeter of the yard:
$$2l + 2w = 2 \cdot 110 + 2 \cdot 85 = 390 ft.$$

$390 \cdot \$12.50 = \$4,875$ to enclose the yard.

TRY IT YOURSELF

77.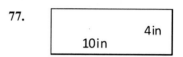

79.

81.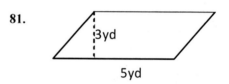

83.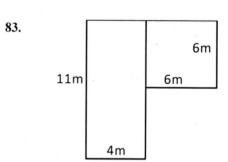

85. $A = bh = 15 \cdot 4 = 60 cm^2$

87. Let x be the length of the third side.
$$22 + 22 + x = 80$$
$$44 + x = 80$$
$$44 + x - 44 = 80 - 44$$
$$x = 36m$$

89. Let x be the length of a side.

$$x + x + x = 85$$
$$3x = 85$$
$$\frac{3x}{3} = \frac{85}{3}$$
$$x = \frac{85}{3} ft = 28\frac{1}{3} ft$$

91. The missing segment has length
$10 - 4 - 4 = 2m$.

$$P = 10 + 6 + 4 + 2 + 2 + 2 + 4 + 6$$
$$= 36m$$

93. $x = 9.1 - 5.4 = 3.7 ft$
$y = 16.3 - 6.2 = 10.1 ft$
$P = 2 \cdot 9.1 + 2 \cdot 16.3$
$= 18.2 + 32.6 = 50.8 ft$

APPLICATIONS

95. The 'upper' and 'lower' sides have $\frac{60}{3} = 20$ trees. The side opposite the house has $\frac{120}{3} = 40$ trees. With the addition of the first tree (as noted in the figure) there are $1 + 20 + 40 + 20 = 81$ trees.

97. One square yard is $3 \cdot 3 = 9 ft^2$.

Ceramic: $\$3.75 / ft^2$

Vinyl: $\frac{\$34.95}{9 ft^2} \approx \$3.88 / ft^2$

The vinyl is more expensive.

99. $A = 14 ft \cdot 20 ft = 280 ft^2$
$280 \cdot \$1.29 = \$361.20$

101. Area of the sail is $A = \frac{1}{2} \cdot 12 ft \cdot 24 ft = 144 ft^2$

One square yard is $3 \cdot 3 = 9 ft^2$

103. $A = \frac{1}{2} h(b_1 + b_2) = \frac{1}{2} \cdot 315 mi (205 mi + 505 mi)$
$= \frac{315 mi}{2}(710 mi) = 111,825 mi^2$

105. Area of the sides:
$48 ft \cdot 12 ft + 48 ft \cdot 12 ft$
$= 576 ft^2 + 576 ft^2 = 1,152 ft^2$

Area of the front and back:
$2(12 ft \cdot 20 ft) = 2(240 ft^2) = 480 ft^2$

Total area = $1,152 ft^2 + 480 ft^2 = 1,632 ft^2$

Each piece of sheetrock covers
$4 ft \cdot 8 ft = 32 ft^2$

$\frac{1,632}{32} = 51$ sheets of sheetrock

WRITING

107. Perimeter is the distance around an object while area is what's enclosed by the object.

109. He put the ^{2} in the wrong place – it should be $25 ft^2$.

REVIEW

111. $8\left(\frac{3}{4}t\right) = \frac{2 \cdot \cancel{8}}{1} \cdot \frac{3}{\cancel{4}} t = 6t$

113. $-\frac{2}{3}(3w - 6) = \left(-\frac{2}{3}\right)3w - \left(-\frac{2}{3}\right)6$
$= -\frac{6w}{3} + \frac{12}{3} = -2w + 4$

115.
$$-\frac{7}{16}x - \frac{3}{16}x = -\frac{10}{16}x$$
$$= -\frac{\cancel{2}\cdot 5}{\cancel{2}\cdot 8}x = -\frac{5}{8}x$$

117.
$$60\left(\frac{3}{20}r - \frac{4}{15}\right) = 60\left(\frac{3}{20}r\right) - 60\left(\frac{4}{15}\right)$$
$$= \cancel{60}^{3}\left(\frac{3}{\cancel{20}}r\right) - \cancel{60}^{4}\left(\frac{4}{\cancel{15}}\right) = 9r - 16$$

Section 9.8: Circles

VOCABULARY

1. A segment drawn from the center of a circle to a point on the circle is called a <u>radius</u>.

3. A <u>diameter</u> is a chord that passes through the center of a circle.

5. The distance around a circle is called its <u>circumference</u>.

7. A diameter of a circle is <u>twice</u> as long as a radius.

CONCEPTS

9. $\overline{OA}, \overline{OC}, \overline{OB}$

11. $\overline{DA}, \overline{DC}, \overline{AC}$

13. ABC, ADC

15. a. Multiply the radius by 2.

    b. Divide the diameter by 2.

17. If C is the circumference of a circle and D is its diameter, then $\frac{C}{D} = \pi$.

19. Square the 6.

NOTATION

21. The symbol $\overset{\frown}{AB}$ is read as "arc AB."

23. a. multiplication: $2 \cdot \pi \cdot r$

    b. raising to a power and multiplication: $\pi \cdot r^2$

GUIDED PRACTICE

25. $C = 2\pi r = 2 \cdot \pi \cdot 4 = 8\pi \, ft \approx 25.1 ft$

27. $C = 2\pi r = 2 \cdot \pi \cdot 6 = 12\pi \, m \approx 37.7 m$

29. The upper portion is a semicircle with radius 6cm.
$$P = \frac{1}{2}(2 \cdot \pi \cdot 6) + 10 + 12 + 10$$
$$= 6\pi + 32 \approx 18.85 + 32 = 50.85 cm$$

31. The upper portion is a semicircle with radius 3m.
$$P = \frac{1}{2}(2 \cdot \pi \cdot 3) + 8 + 8 + 6$$
$$= 3\pi + 22 \approx 9.42 + 22 = 31.42 m$$

33. $D = 6in \rightarrow r = 3in$
$$A = \pi r^2 = \pi \cdot 3^2 = 9\pi \, in^2 \approx 28.3 in^2$$

35. $D = 18in \rightarrow r = 9in$
$$A = \pi r^2 = \pi \cdot 9^2 = 81\pi \, in^2 \approx 254.5 in^2$$

37. The radius of the semicircle is 6cm.
$$A = \frac{1}{2}(\pi \cdot 6^2) + \frac{1}{2} \cdot 12 \cdot 12 = 18\pi + 72$$
$$\approx 56.55 + 72 = 128.6 cm^2$$

39. The radius of the semicircle is 4cm.
$$A = \frac{1}{2}(\pi \cdot 4^2) + 4 \cdot 8 = 8\pi + 32$$
$$\approx 25.1 + 32 = 57.1 cm^2$$

TRY IT YOURSELF

41. Area of rectangle: $4 \cdot 10 = 40 in^2$

    Area of circle: $\pi \cdot 2^2 = 4\pi \, in^2 \approx 12.6 in^2$

    Difference: $40 - 12.6 = 27.4 in^2$

43. Area of parallelogram: $13 \cdot 9 = 117 in^2$

    Area of circle: $\pi \cdot 4^2 = 16\pi \, in^2 \approx 50.3 in^2$

    Difference: $117 - 50.3 = 66.7 in^2$

45. $C = \pi D = \pi \cdot 50 = 50\pi \ yd \approx 157.08 \ yd$

47. If the square has side length 6 inches, the diameter of the circle is 6 inches.
$C = \pi D = 6\pi \ in \approx 18.8 in$

49. If the square has side length 9mm, the radius of the circle is 4.5mm.
$A = \pi r^2 = \pi \cdot 4.5^2 = 20.25\pi \ mm^2 \approx 63.6 mm^2$

APPLICATIONS

51. a. $1 in$

    b. $2 \cdot 1 = 2 in$

    c. $2 \cdot \pi \cdot 1 = 2\pi \ in \approx 2(3.14) = 6.28 in$

    d. $\pi \cdot 1^2 = \pi \ in^2 \approx 3.14 in^2$

53. $D = 2 mi \rightarrow r = 1 mi$
$A = \pi \cdot 1^2 = \pi \ mi^2 \approx 3.14 mi^2$

55. $C = \pi D = 32.66\pi \ ft \approx 32.66(3.14) = 102.6 \ ft$

57. Each lap is $\pi \cdot D = \pi \cdot \dfrac{1}{4} \approx 0.79 mi$.

    $\dfrac{10}{0.79} \approx 12.65$ laps. She needs to run 13 times around the track.

59. Area of the target: $\pi \cdot 2^2 = 4\pi \ ft^2 \approx 12.57 ft^2$

    Area of the bull's eye:
    $\pi \cdot \left(\dfrac{1}{2}\right)^2 = \dfrac{\pi}{4} ft^2 \approx 0.79 ft^2$

    $\dfrac{\frac{\pi}{4}}{4\pi} = \dfrac{\pi}{4} \cdot \dfrac{1}{4\pi} = \dfrac{1}{16} = 0.0625 = 6.25\%$

WRITING

61. The circumference of a circle is the distance around the outside.

63. π is the ratio of a circle's circumference to its diameter.

REVIEW

65. $\dfrac{9}{10} = 0.9 = 90\%$

67. $0.827 = 82.7\%$

69. $\dfrac{\$1.29}{24 oz} = \$0.05375 / oz = 5.375¢ / oz$

71. five

Section 9.9: Volume

VOCABULARY

1. The volume of a three-dimensional figure is a measure of its capacity.

3. cone

5. cylinder

7. pyramid

CONCEPTS

9.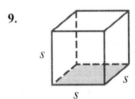

11.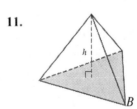

13.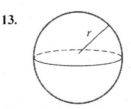

15. cubic inches, mi^3, m^3

17. a. perimeter

    b. volume

    c. area

    d. volume

e. area

f. circumference

19. a. $\dfrac{1}{3}\pi(25)(6) = \dfrac{150}{3}\pi = 50\pi$

 b. $\dfrac{4}{3}\pi(125) = \dfrac{500}{3}\pi$

NOTATION

21. a. cubic inch

 b. $1\,cm^3$

23. a right angle

GUIDED PRACTICE

25. $1\,yd^3 = 1\,yd \cdot 1\,yd \cdot 1\,yd = 3\,ft \cdot 3\,ft \cdot 3\,ft = 27\,ft^3$

27. $1\,km^3 = 1\,km \cdot 1\,km \cdot 1\,km$
 $= 1{,}000m \cdot 1{,}000m \cdot 1{,}000m$
 $= 1{,}000{,}000{,}000\,m^3$

29. $V = lwh = 4 \cdot 2 \cdot 7 = 56\,ft^3$

31. $V = lwh = 5 \cdot 5 \cdot 5 = 125\,in^3$

33. First convert: $0.2m = 20cm$
 $B = \dfrac{1}{2}bh = \dfrac{1}{2} \cdot 3 \cdot 4 = 6\,cm^2$
 $V = Bh = 6 \cdot 20 = 120\,cm^3$

35. First convert: $2\,ft = 24\,in$
 $B = \dfrac{1}{2}bh = \dfrac{1}{2} \cdot 9 \cdot 12 = 54\,in^2$
 $V = Bh = 54 \cdot 24 = 1{,}296\,in^3$

37. $B = 10 \cdot 10 = 100\,yd^2$
 $V = \dfrac{1}{3}Bh = \dfrac{1}{3} \cdot 100 \cdot 21 = 700\,yd^3$

39. $B = 8 \cdot 2 = 16\,ft^2$
 $V = \dfrac{1}{3}Bh = \dfrac{1}{3} \cdot 16 \cdot 6 = 32\,ft^3$

41. $B = \dfrac{1}{2} \cdot 7.2 \cdot 8.3 = 29.88\,ft^2$
 $V = \dfrac{1}{3}Bh = \dfrac{1}{3} \cdot 29.88 \cdot 7 = 69.72\,ft^3$

43. $B = 9\,yd^2$
 $V = \dfrac{1}{3}Bh = \dfrac{1}{3} \cdot 9 \cdot 2 = 6\,yd^3$

45. $V = \pi r^2 h = \pi \cdot 4^2 \cdot 12$
 $= 192\pi\,ft^3 \approx 603.19\,ft^3$

47. $D = 30cm \rightarrow r = 15cm$
 $V = \pi r^2 h = \pi \cdot 15^2 \cdot 14$
 $= 3{,}150\pi\,cm^3 \approx 9{,}896.02\,cm^3$

49. $D = 6m \rightarrow r = 3m$
 $V = \dfrac{1}{3}\pi r^2 h = \dfrac{1}{3} \cdot \pi \cdot 3^2 \cdot 13$
 $= 39\pi\,m^3 \approx 122.52\,m^3$

51. $V = \dfrac{1}{3}\pi r^2 h = \dfrac{1}{3} \cdot \pi \cdot 9^2 \cdot 7$
 $= 189\pi\,yd^3 \approx 593.76\,yd^3$

53. $V = \dfrac{4}{3}\pi r^3 = \dfrac{4}{3} \cdot \pi \cdot 6^3$
 $= 288\pi\,in^3 \approx 904.8\,in^3$

55. $D = 4cm \rightarrow r = 2cm$
 $V = \dfrac{4}{3}\pi r^3 = \dfrac{4}{3} \cdot \pi \cdot 2^3$
 $= \dfrac{32}{3}\pi\,cm^3 \approx 33.5\,cm^3$

TRY IT YOURSELF

57. A sphere would be
 $V = \dfrac{4}{3}\pi r^3 = \dfrac{4}{3} \cdot \pi \cdot 9^3 = 972\pi\,ft^3$ so a hemisphere would have volume
 $\dfrac{972\pi}{2} = 486\pi\,in^3 \approx 1{,}526.81\,in^3$.

59. $V = \pi r^2 h = \pi \cdot 6^2 \cdot 12$
$= 432\pi \, m^3 \approx 1{,}357.17 m^3$

61. $V = lwh = 3 \cdot 4 \cdot 5 = 60 cm^3$

63. $D = 10 cm \rightarrow r = 5 cm$
$V = \frac{1}{3}\pi r^2 h = \frac{1}{3} \cdot \pi \cdot 5^2 \cdot 12$
$= 100\pi \, cm^3 \approx 314.16 cm^3$

65. $B = 10 \cdot 10 = 100 m^2$
$V = \frac{1}{3}Bh = \frac{1}{3} \cdot 100 \cdot 12 = 400 m^3$

67. $B = \frac{1}{2}bh = \frac{1}{2} \cdot 3 \cdot 4 = 6 m^2$
$V = Bh = 6 \cdot 8 = 48 m^3$

69. Break the figure down into a rectangular solid and a pyramid.

Rectangular solid: $V = 8 \cdot 8 \cdot 8 = 512 cm^3$

Pyramid: $V = \frac{1}{3}Bh = \frac{1}{3}(8 \cdot 8)3 = 64 cm^3$

Total volume: $512 + 64 = 576 cm^3$

71. Note that the outsides of the cylinder form a sphere.

Cylinder: $V = \pi r^2 h = \pi \cdot 3^2 \cdot 16 = 144\pi \, cm^3$

Sphere: $V = \frac{4}{3}\pi r^3 = \frac{4}{3} \cdot \pi \cdot 3^3 = 36\pi$

Total volume:
$144\pi + 36\pi = 180\pi \, cm^3 \approx 565.49 cm^3$

APPLICATIONS

73. $V = s^3 = \left(\frac{1}{2}\right)^3 = \frac{1}{8} in^3 = 0.125 in^3$

75. Begin by converting everything to feet:
$V = (27")(17")(8") = \frac{27}{12} \cdot \frac{17}{12} \cdot \frac{8}{12}$
$= \frac{9}{4} \cdot \frac{17}{12} \cdot \frac{2}{3} = \frac{306}{144} = 2.125$

77. $D = 6 ft \rightarrow r = 3 ft$
$V = \pi r^2 h = \pi \cdot 3^2 \cdot 7 = \pi \cdot 9 \cdot 7 = 63\pi \, ft^3$
$63\pi \approx 197.92 ft^3$

79. $D = 40 ft \rightarrow r = 20 ft$
$V = \frac{4}{3}\pi r^3 = \frac{4}{3}\pi(20)^3 = \frac{32{,}000}{3}\pi \, ft^3$
$\frac{32{,}000}{3}\pi \approx 33{,}510.32 ft^3$

81. $\frac{30.4}{3.8} = \frac{304}{38} = \frac{2 \cdot 2 \cdot 2 \cdot 2 \cdot 19}{2 \cdot 19} = \frac{8}{1}$
$8 : 1$

83. a. the radius of the hemisphere is 15 inches.
$V = \frac{1}{2}\left(\frac{4}{3}\pi r^3\right) = \frac{2}{3}\pi(15^3) = 2{,}250\pi \, in^3$
$2{,}250\pi \, in^3 \approx 7{,}068.58 in^3$

b. $\frac{7{,}068.58}{231} \approx 30.6 \, gal$

WRITING

85. Volume is the measure of capacity of a figure.

87. Yes, area is a two dimensional measurement while volume is a three dimensional measurement.

REVIEW

89. $-5(5-2)^2 + 3 = -5(3)^2 + 3 = -5 \cdot 9 + 3$
$= -45 + 3 = -42$

91. $-x = 4$
$\frac{-x}{-1} = \frac{4}{-1}$
$x = -4$

93. $\dfrac{3in}{15in} = \dfrac{\cancel{3}\cancel{in}}{\cancel{3}\cdot 5\cancel{in}} = \dfrac{1}{5}$

95. $\dfrac{2.4m}{1}\cdot\dfrac{1,000mm}{1m} = 2,400mm$

Chapter 9 Review

1. point: C or D

   line: CD

   plane: GHI

3. $\angle ABC$; $\angle CBA$; $\angle B$; $\angle 1$

5. acute: $\angle 1$, $\angle 2$

   right: $\angle ABD$, $\angle CBD$

   obtuse: $\angle CBE$

   straight: $\angle ABC$

7. Yes – they have the same origin and direction.

9. $x + 35° = 50°$
   $x + 35° - 35° = 50° - 35°$
   $x = 15°$

11. a. vertical angles:

    $m(\angle 1) = 65°$

    b. supplemental angles:

    $m(\angle 2) + 65° = 180°$
    $m(\angle 2) + 65° - 65° = 180° - 65°$
    $m(\angle 2) = 115°$

13. a. vertical angles:

    $5x + 25° = 6x + 5°$
    $5x + 25° - 5° = 6x + 5° - 5°$
    $5x + 20° = 6x$
    $5x + 20° - 5x = 6x - 5x$
    $20° = x$

    b. substitute x:

    $m(\angle HFI) = 6(20°) + 5° = 120° + 5° = 125°$

    c. supplemental angles:

    $m(\angle GFH) + 125° = 180°$
    $m(\angle GFH) + 125° - 125° = 180° - 125°$
    $m(\angle GFH) = 55°$

15. $180° - 143° = 37°$

17. a. parallel

    b. transversal

    c. perpendicular

19. $\angle 1 \,\&\, \angle 5$; $\angle 4 \,\&\, \angle 8$
    $\angle 2 \,\&\, \angle 6$; $\angle 3 \,\&\, \angle 7$

21. $m(\angle 2) = m(\angle 4) = m(\angle 6) = 110°$
    $m(\angle 1) + 110° = 180°$
    $m(\angle 1) + 110° - 110° = 180° - 110°$
    $m(\angle 1) = 70°$
    $m(\angle 1) = m(\angle 3) = m(\angle 5) = m(\angle 7) = 70°$

23. a. alternate interior angles:

    $2x - 30° = x + 10°$
    $2x - 30° + 30° = x + 10° + 30°$
    $2x = x + 40°$
    $2x - x = x + 40° - x$
    $x = 40°$

    b. both are $40° + 10° = 50°$

25. a. The angles are congruent:

    $7x - 46° = 2x + 9°$
    $7x - 46° + 46° = 2x + 9° + 46°$
    $7x = 2x + 55°$
    $7x - 2x = 2x + 55° - 2x$
    $5x = 55°$
    $\dfrac{5x}{5} = \dfrac{55°}{5}$
    $x = 11°$

b. both are $2(11°)+9°=22°+9°=31°$

27. a. 8, octagon, 8

b. 5, pentagon, 5

c. 3, triangle, 3

d. 6, hexagon, 6

e. 4, quadrilateral, 4

f. 10, decagon, 10

29. a. acute

b. right

c. obtuse

d. acute

31. $70°+20°+x=180°$
$90°+x=180°$
$90°+x-90°=180°-90°$
$x=90°$

33. $m(\angle A)+32°+77°=180°$
$m(\angle A)+109°=180°$
$m(\angle A)+109°-109°=180°-109°$
$m(\angle A)=71°$

35. $2(65°)+x=180°$
$130°+x=180°$
$130°+x-130°=180°-130°$
$x=50°$

37. $2(56.5°)+m(\angle C)=180°$
$113°+m(\angle C)=180°$
$113°+m(\angle C)-113°=180°-113°$
$m(\angle C)=67°$

39. $a^2+b^2=c^2$
$5^2+12^2=c^2$
$25+144=c^2$
$169=c^2$
$\sqrt{169}=c$
$13 cm=c$

41. $a^2+b^2=c^2$
$a^2+77^2=85^2$
$a^2+5,929=7,225$
$a^2+5,929-5,929=7,225-5,929$
$a^2=1,296$
$a=\sqrt{1,296}$
$a=36 in$

43. $a^2+5^2=16^2$
$a^2+25=256$
$a^2+25-25=256-25$
$a^2=231$
$a=\sqrt{231}m \approx 15.20m$

45. $48^2+55^2=c^2$
$2,304+3,025=c^2$
$5,329=c^2$
$\sqrt{5,329}=c$
$73 in=c$

47. $8^2+11^2\stackrel{?}{=}15^2$
$64+121\stackrel{?}{=}225$
$185=225$
Not a right triangle.

49. a. $\angle D$

b. $\angle E$

c. $\angle F$

d. $\overline{DF}$

e. $\overline{DE}$

f. $\overline{EF}$

51. congruent: SSS

53. not necessarily congruent

55. yes

57.
$\dfrac{x}{16} = \dfrac{8}{32}$ $\dfrac{y}{7} = \dfrac{32}{8}$

$32 \cdot x = 8 \cdot 16$ $8 \cdot y = 32 \cdot 7$

$\dfrac{32x}{32} = \dfrac{128}{32}$ $\dfrac{8y}{8} = \dfrac{224}{8}$

$x = 4$ $y = 28$

59.
a. trapezoid

b. square

c. parallelogram

d. rectangle

e. rhombus

f. rectangle

61.
a. true

b. true

c. true

d. false

63.
$S = (n-2)180°$

$S = (8-2)180°$

$S = 6 \cdot 180°$

$S = 1,080°$

65. $P = 4l = 4 \cdot 18 = 72 \, in$

67. $8 + 6 + 8 + 4 + 4 = 30 \, m$

69.
$2(24) + b = 107$

$48 + b = 107$

$48 + b - 48 = 107 - 48$

$b = 59 \, ft$

71. $A = s^2 = 3.1^2 = 9.61 \, cm^2$

73. $A = b \cdot h = 30 \cdot 15 = 450 \, ft^2$

75.
$A = \dfrac{1}{2} h(b_1 + b_2) = \dfrac{1}{2} \cdot 8(12 + 18)$

$= 4(30) = 120 \, cm^2$

77.
Trapezoid: $\dfrac{1}{2} \cdot 8(12 + 20) = 4(32) = 128 \, ft^2$

Triangle: $\dfrac{1}{2} \cdot 12 \cdot 4 = 6 \cdot 4 = 24 \, ft^2$

Area of figure: $128 + 24 = 152 \, ft^2$

79.
$A = bh$

$240 = 30 \cdot h$

$\dfrac{240}{30} = \dfrac{30h}{30}$

$8 \, ft = h$

81.
$P = 2l + 2w = 2 \cdot 115 + 2 \cdot 78$

$= 230 + 156 = 386 \, ft$

$386(8.5) = \$3,281$

83.
a. $\overline{CD}$, $\overline{AB}$

b. $\overline{AB}$

c. $\overline{OA}$, $\overline{OC}$, $\overline{OD}$, $\overline{OB}$

d. O

85. The outer circle has circumference $C = \pi d = 8\pi \, cm \approx 25.1 \, cm$. The total perimeter, then, is approximately $25.1 + 10 + 10 = 45.1 \, cm$.

87. The area of the circle (which has radius 4 centimeters) is $A = \pi r^2 = \pi \cdot 4^2 = 16\pi \, cm^2 \approx 50.3 \, cm^2$. The remaining rectangle has area $l \cdot w = 8 \cdot 10 = 80 \, cm^2$. The total area of the figure is approximately $80 + 50.3 = 130.3 \, cm^2$.

89. $V = s^3 = 5^3 = 125 \, cm^3$

93. $V = \frac{1}{3}\pi r^2 h = \frac{1}{3}\pi \cdot 5^2 \cdot 30$
$= 250\pi \, in^3 \approx 785.40 \, in^3$

95. $B = \frac{1}{2} \cdot 12 \cdot 35 = 210 m^2$
$A = \frac{1}{3} Bh = \frac{1}{3} \cdot 210 \cdot 42 = 2,940 m^3$

97. Combine half a sphere (of radius 5 feet) with a cylinder.
$V = \frac{1}{2} \cdot \frac{4}{3}\pi \cdot 5^3 + \pi \cdot 5^2 \cdot 16$
$= \frac{250}{3}\pi + 400\pi$
$\approx 1,518 \, ft^3$

99. $1 ft = 12 in$
$1 ft^3 = (12 in)^3 = 12^3 in^3 = 1,728 in^3$

Chapter 9 Test

1. a. $135°$; obtuse
   b. $90°$; right
   c. $40°$; acute
   d. $180°$; straight

3. D

5. Vertical angles have the same measure:
$3x = 2x + 20°$
$3x - 2x = 2x + 20° - 2x$
$x = 20°$
$m(\angle ABD) = m(\angle ABD)$
$= 3x = 3(20°) = 60°$

7. a. l_2 intersects two coplanar lines. It is called a <u>transversal</u>.

   b. $\angle 4$ and $\angle 6$ are alternate interior angles.

   c. $\angle 3$ and $\angle 7$ are corresponding angles.

9. The two angles sum to $180°$.
$x + 20° + 2x + 10° = 180°$
$3x + 30° = 180°$
$3x + 30° - 30° = 180° - 30°$
$3x = 150°$
$\frac{3x}{3} = \frac{150°}{3}$
$x = 50°$
$x + 20° = 50° + 20° = 70°$
$2x + 10° = 2(50°) + 10° = 110°$

11. a. isosceles
    b. scalene
    c. equilateral
    d. isosceles

13. $2b + 12° = 180°$
$2b + 12° - 12° = 180° - 12°$
$2b = 168°$
$\frac{2b}{2} = \frac{168°}{2}$
$b = 84°$

15. a. $m(\overline{RS}) = m(\overline{QT}) = 10$
    b. $x = m(\angle T) = 65°$
    c. $y = 180° - x = 180° - 65° = 115°$
    d. $z = y = 115°$

17. $25 + 36 + 37 + 48 + 42 = 188 in$

19. Area of rectangle: $lw = 25 \cdot 16 = 400 cm^2$

   Area of triangle: $\frac{1}{2}bh = \frac{1}{2} \cdot 10 \cdot 8 = 40 cm^2$

   Area shaded: $400 - 40 = 360 cm^2$

21. $1 ft = 12 in \rightarrow 1 ft^2 = (12 in)^2 = 12^2 in^2 = 144 in^2$

23. a. $\overline{RS}$, $\overline{XY}$

   b. $\overline{XY}$

   c. $\overline{OX}$, $\overline{OR}$, $\overline{OS}$, $\overline{OY}$

25. $C = \pi d = \pi \cdot 21 = 21\pi\ ft \approx 66.0\ ft$

27. Since the diameter is 30 meters, the radius is 15 meters.
   $A = \pi r^2 = \pi \cdot 15^2 = 225\pi\ m^2 \approx 706.9 m^2$.

29. a. congruent: SSS

   b. congruent: ASA

   c. not necessarily congruent

   d. congruent: SAS

31. a. yes b. yes

33. $\frac{t}{7} = \frac{6}{2}$

   $t \cdot 2 = 6 \cdot 7$

   $\frac{2t}{2} = \frac{42}{2}$

   $t = 21 ft$

35. $25^2 + 19^2 = d^2$

   $625 + 361 = d^2$

   $986 = d^2$

   $\sqrt{986} in = d \approx 31.4 in$

37. $V = s^3 = 6^3 = 216 m^3$

39. $V = \frac{1}{3}\pi r^2 h = \frac{1}{3}\pi \cdot 12^2 \cdot 27$
   $= 1,296\pi\ in^3 \approx 4,071.50 in^3$

41. $B = \frac{1}{2} \cdot 20 \cdot 21 = 210\ ft^2$

   $V = \frac{1}{3} Bh = \frac{1}{3} \cdot 210 \cdot 27 = 1,890\ ft^3$

43. $B = 10 \cdot 10 = 100\ yd^2$

   $V = \frac{1}{3} Bh = \frac{1}{3} \cdot 100 \cdot 12 = 400\ mi^3$

45. Combine half a sphere (of radius 15 feet) with a cylinder.

   $V = \frac{1}{2} \cdot \frac{4}{3}\pi \cdot 15^3 + \pi \cdot 15^2 \cdot 40$
   $= 2,250\pi + 9,000\pi$
   $= 11,250\pi\ ft^3 \approx 35,343\ ft^3$

Chapters 1 – 9 Cumulative Review

1. The best offer was $8,995.

3. $^{1\ 1\ 2\ 3}458$
   $8,099$
   $23,419$
   $+58$
   $\overline{32,034}$

5. a. $P = 204 + 97 + 204 + 97 = 602\ ft$

   b. $A = 204 \cdot 97 = 19,788\ ft^2$

7. 2 coats means $2 \cdot 8,400 = 16,800\ ft^2$

   48
   $350\overline{)16800}$
   -1400
   2800
   -2800
   0

   48 gallons are needed.

9. $16 = 2 \cdot 2 \cdot 2 \cdot 2$
    $24 = 2 \cdot 2 \cdot 2 \cdot 3$
    $LCM(16, 24) = 2 \cdot 2 \cdot 2 \cdot 2 \cdot 3 = 48$
    $GCF(16, 24) = 2 \cdot 2 \cdot 2 = 8$

11. a. $\{..., -2, -1, 0, 1, 2, ...\}$

    b. $-(-3) = 3$

13. $30 - (55 + 75 + 20 + 20) = 30 - 170 = -\$140$

15. a. $\dfrac{35}{28} = \dfrac{5 \cdot \cancel{7}}{4 \cdot \cancel{7}} = \dfrac{5}{4}$

    b. $\dfrac{3}{8} = \dfrac{3}{8} \cdot \dfrac{6}{6} = \dfrac{18}{48}$

    c. $\dfrac{8}{9}$

    d. $7\dfrac{1}{2} = \dfrac{2 \cdot 7 + 1}{2} = \dfrac{15}{2}$

17. $-\dfrac{5}{77}\left(\dfrac{33}{50}\right) = -\dfrac{\cancel{5} \cdot 3 \cdot \cancel{11}}{7 \cdot \cancel{11} \cdot \cancel{5} \cdot 10}$
    $= -\dfrac{3}{7 \cdot 10} = -\dfrac{3}{70}$

19. $\dfrac{3}{4} - \dfrac{3}{5} = \dfrac{3}{4} \cdot \dfrac{5}{5} - \dfrac{3}{5} \cdot \dfrac{4}{4} = \dfrac{15}{20} - \dfrac{12}{20} = \dfrac{3}{20}$

21. $45\dfrac{2}{3} + 96\dfrac{4}{5} = 45\dfrac{10}{15} + 96\dfrac{12}{15} = 141\dfrac{22}{15}$
    $= 141 + \dfrac{15}{15} + \dfrac{7}{15} = 141 + 1 + \dfrac{7}{15}$
    $= 142\dfrac{7}{15}$

23. $\dfrac{1}{8} \cdot \dfrac{3}{4} = \dfrac{3}{32}\ fl\ oz$

25. $\dfrac{3}{4} + \left(-\dfrac{1}{3}\right)^2 \left(\dfrac{5}{4}\right) = \dfrac{3}{4} + \dfrac{1}{9} \cdot \dfrac{5}{4} = \dfrac{3}{4} + \dfrac{5}{36}$
    $= \dfrac{3}{4} \cdot \dfrac{9}{9} + \dfrac{5}{36} = \dfrac{27}{36} + \dfrac{5}{36} = \dfrac{32}{36}$
    $= \dfrac{\cancel{4} \cdot 8}{\cancel{4} \cdot 9} = \dfrac{8}{9}$

27. $\overset{1\ 1}{}3.400$
    106.780
    35.000
    $+0.008$
    $\overline{145.188}$

29. $-8.8 + (-7.3 - 9.5)$
    $= -8.8 + (-16.8)$
    $= -25.6$

31. $\dfrac{0.0742}{1.4} = \dfrac{0.742}{14} = 0.053$

33. $2.17(1,250) = \$2,712.50$

35. $18,000 \div 9 = 2,000$

37. $15\overline{)2.000}0.133$
    $\underline{-15}$
    50
    $\underline{-45}$
    50
    $\underline{-45}$
    5

    $\dfrac{2}{15} = 0.1\overline{3}$

39.

41. $\dfrac{\$24}{400 in^2} = \$0.06/in^2$

$\dfrac{\$42}{600 in^2} = \$0.07/in^2$

The smaller board is a better buy.

43. $\dfrac{3}{1,000} = \dfrac{375}{x}$

$3 \cdot x = 375 \cdot 1,000$

$\dfrac{3x}{3} = \dfrac{375,000}{3}$

$x = 125,000$

45.
a. $\dfrac{168 in}{1} \cdot \dfrac{1 ft}{12 in} = \dfrac{168}{12} ft = 14 ft$

b. $\dfrac{212 oz}{1} \cdot \dfrac{1 lb}{16 oz} = \dfrac{212}{16} lb = 13.25 lb$

c. $\dfrac{30g}{1} \cdot \dfrac{4qt}{1g} = 30 \cdot 4qt = 120qt$

d. $\dfrac{12.5h}{1} \cdot \dfrac{60 \min}{1h} = 12.5(60) \min = 750 \min$

47. $\dfrac{240,000m}{1} \cdot \dfrac{1km}{1,000m} = \dfrac{240,000}{1,000} km = 240 km$

49. $\dfrac{10 lb}{1} \cdot \dfrac{1 kg}{2.2 lb} = \dfrac{10}{2.2} kg \approx 4.5 kg$

51.

Percent	Decimal	Fraction
57%	0.57	$\dfrac{57}{100}$
0.1%	0.001	$\dfrac{1}{1,000}$
$33.333\% = 33\dfrac{1}{3}\%$	0.3333...	$\dfrac{1}{3}$

53. $x = 0.15 \cdot 450$

$x = 67.5$

55. $51 = x \cdot 60$

$\dfrac{51}{60} = \dfrac{60x}{60}$

$0.85 = x$

51 is 85% of 60.

57. $x = 0.0625 \cdot 18,550 = \$1,159.38$

59. Use $140 as a base price. 10% of $140 is $14. Half that is $7. $14+$7 = $21 tip.

61.
a. about 7.5 little cars, so $7.5(50,000) = 375,000$ cars.

b. 6 little cars, so $6(50,000) = 300,000$ cars.

c. Seattle: $6.1(50,000) = 305,000$ cars

Minneapolis: $4.25(50,000) = 212,500$

Difference: $305,000 - 212,500 = 92,500$ cars.

63.
a. food, about $17.5 billion

b. about $2.2 billion

c. about $12.2 – $3.8 = $8.4 billion

65. $\dfrac{2x+3y}{z-y}$

$\dfrac{2(2)+3(-3)}{(-4)-(-3)} = \dfrac{4+(-9)}{-4+3}$

$= \dfrac{-5}{-1} = 5$

67.
a. $12(4a) = (12 \cdot 4)a = 48a$

b. $-2b(-7)(3) = (-2(-7)3)b = 42b$

69.
a. $10x - 7x = 3x$

b. $c^2 + 4c^2 + 2c^2 - c^2$
$= 5c^2 + c^2 = 6c^2$

c. $4m - n - 12m + 7n = -8m + 6n$

$$4x-2(3x-4)-5(2x)$$
d. $= 4x-6x+8-10x$
$= -12x+8$

71. $\dfrac{x}{8}-2=-5$

$\dfrac{x}{8}-2+2=-5+2$

$\dfrac{x}{8}=-3$

$8\cdot\dfrac{x}{8}=8(-3)$

$x=-24$

Check:

$\dfrac{x}{8}-2=-5$

$\dfrac{-24}{8}-2\overset{?}{=}-5$

$-3-2\overset{?}{=}-5$

$-5=-5$

The solution checks.

73. $3(2p+15)=3p-4(11-p)$

$6p+45=3p-44+4p$

$6p+45=7p-44$

$6p+45+44=7p-44+44$

$6p+89=7p$

$6p+89-6p=7p-6p$

$89=p$

Check:

$3(2p+15)=3p-4(11-p)$

$3(2\cdot 89+15)\overset{?}{=}3\cdot 89-4(11-89)$

$3(178+15)\overset{?}{=}267-4(-78)$

$3(193)\overset{?}{=}267+312$

$579=579$

The solution checks.

75. Let x be the number of visits remaining.

$48+6x=90$

$48+6x-48=90-48$

$6x=42$

$\dfrac{6x}{6}=\dfrac{42}{6}$

$x=7$

She must make 7 more visits.

77. a. $s^4\cdot s^5\cdot s = s^{4+5+1} = s^{10}$

b. $\left(a^5\right)^7 = a^{35}$

c. $\left(r^2 t^4\right)\left(r^3 t^5\right) = r^{2+3}t^{4+5} = r^5 t^9$

d. $\left(2b^3 c^6\right)^3 = 2^3\left(b^3\right)^3\left(c^6\right)^3 = 8b^9 c^{18}$

e. $\left(y^5\right)^2\left(y^4\right)^3 = y^{10}y^{12} = y^{22}$

f. $\left[(-5.5)^3\right]^{12} = (-5.5)^{36}$

79. a.

$x+105°=180°$

$x+105°-105°=180°-105°$

$x=75°$

b.

$x+75°=90°$

$x+75°-75°=90°-75°$

$x=15°$

81. Notice that it is an isosceles trapezoid:

a. alternate interior angles.

$m(\angle 1)=75°$

b. isosceles triangle:

$2 \cdot 75° + m(\angle C) = 180°$

$150° + m(\angle C) = 180°$

$150° + m(\angle C) - 150° = 180° - 150°$

$m(\angle C) = 30°$

c. supplemental angles:

$180° - m(\angle 1) = 180° - 75° = 105° = m(\angle 2)$

d.

$m(\angle 3) + 75° = 180°$

$m(\angle 3) + 75° - 75° = 180° - 75°$

$m(\angle 3) = 105°$

83. $34° + 2x = 180°$

$34° + 2x - 34° = 180° - 34°$

$2x = 146°$

$\dfrac{2x}{2} = \dfrac{146°}{2}$

$x = 73°$

85. $16^2 + 63^2 \stackrel{?}{=} 65^2$

$256 + 3{,}969 \stackrel{?}{=} 4{,}225$

$4{,}225 = 4{,}225$

It is a right triangle.

87. $S = (n-2)180°$

$S = (5-2)180°$

$S = 3 \cdot 180°$

$S = 540°$

89. $A = \dfrac{1}{2}bh = \dfrac{1}{2} \cdot 14 \cdot 18 = 7 \cdot 18 = 126\,ft^2$

91. $1\,ft = 12\,in \rightarrow 1\,ft^2 = (12\,in)^2 = 12^2\,in^2 = 144\,in^2$

93. The two semicircles combine to form a single circle with radius $9.6\,yd$. The area of the 'missing' part from the semicircles is $\pi \cdot 9.6^2 \approx 289.53\,yd^2$. The area of the rectangle without the missing parts is $20.2 \cdot 19.2 = 387.84\,yd^2$. The shaded area, then, is roughly $387.84 - 289.53 = 98.31\,yd^2$.

95. If the diameter is 18 inches, the radius is 9 inches.

$V = \dfrac{4}{3}\pi r^3 = \dfrac{4}{3}\pi \cdot 9^3 = \dfrac{4}{3}\pi \cdot 729$

$= 972\pi\,in^3 \approx 3{,}053.63\,in^3$

97. $V = \pi r^2 h = \pi \cdot 1^2 \cdot 20 = 20\pi\,ft^3 \approx 62.83\,ft^3$

Appendices

Section II.1: Introduction to Polynomials

VOCABULARY

1. A polynomial with one term is called a <u>monomial</u>.

3. A polynomial with two terms is called a <u>binomial</u>.

CONCEPTS

5. binomial

7. monomial

9. monomial

11. trinomial

13. 3

15. 2

17. 1

19. 7

NOTATION

21. $3a^2 + 2a - 7 = 3(2)^2 + 2(2) - 7$
$= 3(4) + 4 - 7$
$= 12 + 4 - 7$
$= 16 - 7$
$= 9$

PRACTICE

23. $3x + 4$
$3(3) + 4 = 9 + 4 = 13$

25. $2x^2 + 4$
$2(-1)^2 + 4 = 2(1) + 4 = 2 + 4 = 6$

27. $0.5t^3 - 1$
$0.5 \cdot 4^3 - 1 = 0.5 \cdot 64 - 1 = 32 - 1 = 31$

29. $\frac{2}{3}b^2 - b + 1$
$\frac{2}{3} \cdot 3^2 - 3 + 1 = 6 - 3 + 1 = 4$

31. $-2s^2 - 2s + 1$
$-2(-1)^2 - 2(-1) + 1 = -2 \cdot 1 + 2 + 1$
$= -2 + 2 + 1 = 0 + 1 = 1$

APPLICATIONS

33. $h = -16t^2 + 64t$
$= -16 \cdot 0^2 + 64 \cdot 0$
$= -0 + 0 = 0 \, ft$

35. $h = -16t^2 + 64t$
$= -16 \cdot 2^2 + 64 \cdot 2$
$= -64 + 128 = 64 \, ft$

37. $d = 0.04v^2 + 0.9v$
$= 0.04 \cdot 30^2 + 0.9 \cdot 30$
$= 36 + 27 = 63 \, ft$

39. $d = 0.04v^2 + 0.9v$
$= 0.04 \cdot 60^2 + 0.9 \cdot 60$
$= 144 + 54 = 198 \, ft$

WRITING

41. To find the degree of the polynomial look at the degrees of the individual terms and then take the largest of those numbers.

REVIEW

43. $\frac{2}{3} + \frac{4}{3} = \frac{6}{3} = 2$

45. $\frac{5}{12} \cdot \frac{18}{5} = \frac{\cancel{5} \cdot 3 \cdot \cancel{6}}{2 \cdot \cancel{6} \cdot \cancel{5}} = \frac{3}{2} = 1\frac{1}{2}$

47. $x - 4 = 12$
$x - 4 + 4 = 12 + 4$
$x = 16$

49. $2(x-3) = 6$

$2x - 6 = 6$

$2x - 6 + 6 = 6 + 6$

$2x = 12$

$\dfrac{2x}{2} = \dfrac{12}{2}$

$x = 6$

Section II.2: Adding and Subtracting Polynomials

VOCABULARY

1. If two algebraic terms have exactly the same variables and exponents, they are called <u>like</u> terms.

CONCEPTS

3. To add two monomials, we add the <u>coefficients</u> and keep the same <u>variables</u> and exponents.

5. yes, $3y + 4y = 7y$

7. no

9. yes, $3x^3 + 4x^3 + 6x^3 = 13x^3$

11. yes, $-5x^2 + 13x^2 + 7x^2 = 15x^2$

NOTATION

13. $(3x^2 + 2x - 5) + (2x^2 - 7x)$

$= (3x^2 + 2x^2) + (2x - 7x) + (5)$

$= 5x^2 + (-5x) + 5$

$= 5x^2 - 5x + 5$

PRACTICE

15. $4y + 5y = 9y$

17. $8t^2 + 4t^2 = 12t^2$

19. $3s^2 + 4s^2 + 7s^2 = 14s^2$

21. $\dfrac{1}{8}a + \dfrac{3}{8}a + \dfrac{5}{8}a = \dfrac{1+3+5}{8}a = \dfrac{9}{8}a$

23. $\dfrac{2}{3}c^2 + \dfrac{1}{3}c^2 + \dfrac{2}{3}c^2 = \dfrac{2+1+2}{3}c^2 = \dfrac{5}{3}c^2$

25. $(3x + 7) + (4x - 3)$

$= 7x + 4$

27. $(2x^2 + 3) + (5x^2 - 10)$

$= 7x^2 - 7$

29. $(5x^3 - 42x) + (7x^3 - 107x)$

$= 12x^3 - 149x$

31. $(3x^2 + 2x - 4) + (5x^2 - 17)$

$= 8x^2 + 2x - 21$

33. $(7y^2 + 5y) + (y^2 - y - 2)$

$= 8y^2 + 4y - 2$

35. $(3x^2 - 3x - 2) + (3x^2 + 4x - 3)$

$= 6x^2 + x - 5$

37. $(2.5a^2 + 3a - 9) + (3.6a^2 + 7a - 10)$

$= 6.1a^2 + 10a - 19$

39. $(3n^2 - 5.8n + 7) + (-n^2 + 5.8n - 2)$

$= 2n^2 + 5$

41. $\begin{array}{r} 3x^2 + 4x + 5 \\ +2x^2 - 3x + 6 \\ \hline 5x^2 + x + 11 \end{array}$

43. $\begin{array}{r} -3x^2 - 7 \\ +\; -4x^2 - 5x + 6 \\ \hline -7x^2 - 5x - 1 \end{array}$

45. $\begin{array}{r} -3x^2 + 4x + 25.4 \\ +\; 5x^2 - 3x - 12.5 \\ \hline 2x^2 + x + 12.9 \end{array}$

47. $32u^3 - 16u^3 = 16u^3$

49. $18x^5 - 11x^5 = 7x^5$

51. $(30x^2-4)-(11x^2+1)$
$=30x^2-4-11x^2-1$
$=19x^2-5$

53. $(3x^2-2x-1)-(-4x^2+4)$
$=3x^2-2x-1+4x^2-4$
$=7x^2-2x-5$

55. $(4.5a+3.7)-(2.9a-4.3)$
$=4.5a+3.7-2.9a+4.3$
$=1.6a+8$

57. $(2b^2+3b-5)-(2b^2-4b-9)$
$=2b^2+3b-5-2b^2+4b+9$
$=7b+4$

59. $(5p^2-p+71)-(4p^2+p+71)$
$=5p^2-p+71-4p^2-p-71$
$=p^2-2p$

61. $(3.7y^2-5)-(2y^2-3.1y+4)$
$=3.7y^2-5-2y^2+3.1y-4$
$=1.7y^2+3.1y-9$

63. $\quad 3x^2+4x-5$
$\quad -(-2x^2-2x+3)$
$\quad \overline{\quad 5x^2+6x-8\quad}$

65. $\quad -2x^2-4x+12$
$\quad -\ (10x^2+9x-24)$
$\quad \overline{-12x^2-13x+36}$

67. $\quad 4x^3-3x+10$
$\quad -\ (5x^3-4x-4)$
$\quad \overline{-x^3\ +\ x+14}$

APPLICATIONS

69. Each 'bulge' has 3 sides, and there are 4 bulges. If each side has length x, the perimeter is $12x$.

71. $2(x+4)+2x=2x+8+2x=(4x+8)\,ft$

WRITING

73. Like terms are terms where the variables and the exponents both match.

75. To subtract two polynomials, add the opposite of the coefficients of any like terms from the subtracted polynomial.

REVIEW

77. $14.6-13.8=0.8\,oz$

79. Let x be the amount it still needs to be lowered.

$85-31+x=0$
$54+x=0$
$54+x-54=0-54$
$x=-54$

It needs to be lowered an additional 54 feet.

Section II.3: Multiplying Polynomials

VOCABULARY

1. $(2x^3)(3x^4)$ is the product of two monomials.

3. In the acronym FOIL, F stands for first terms, O for outer terms, I for inner terms, and L for last terms.

CONCEPTS

5. To multiply two polynomials, multiply each term of one polynomial by each term of the other polynomial, and then combine like terms.

7. a. $6x^2+x-12$

   b. $5x^4+8ax^2+3a^2$

NOTATION

9. $(9n^3)(8n^2)=(9\cdot 8)(n^3\cdot n^2)=72n^5$

11. $(2x+5)(3x-2)$
$= 2x(3x) - 2x(2) + 5(3x) - 5(2)$
$= 6x^2 - 4x + 15x - 10$
$= 6x^2 + 11x - 10$

PRACTICE

13. $(3x^2)(4x^3) = 12x^5$

15. $(3b^2)(-2b) = -6b^3$

17. $(-2x^2)(3x^3) = -6x^5$

19. $\left(-\dfrac{2}{3}y^5\right)\left(\dfrac{3}{4}y^2\right) = -\dfrac{6}{12}y^7 = -\dfrac{1}{2}y^7$

21. $3(x+4) = 3 \cdot x + 3 \cdot 4 = 3x + 12$

23. $-4(t+7) = -4 \cdot t + (-4)7 = -4t - 28$

25. $3x(x-2) = 3x \cdot x - 3x \cdot 2 = 3x^2 - 6x$

27. $-2x^2(3x^2 - x)$
$= (-2x^2)3x^2 - (-2x^2)x$
$= -6x^4 + 2x^3$

29. $2x(3x^2 + 4x - 7)$
$= 2x \cdot 3x^2 + 2x \cdot 4x - 2x \cdot 7$
$= 6x^3 + 8x^2 - 14x$

31. $-p(2p^2 - 3p + 2)$
$= -p(2p^2) - (-p)3p + (-p)2$
$= -2p^3 + 3p^2 - 2p$

33. $3q^2(q^2 - 2q + 7)$
$= 3q^2 \cdot q^2 - 3q^2 \cdot 2q + 3q^2 \cdot 7$
$= 3q^4 - 6q^3 + 21q^2$

35. $(a+4)(a+5)$
$= a^2 + 5a + 4a + 20$
$= a^2 + 9a + 20$

37. $(3x-2)(x+4)$
$= 3x^2 + 12x - 2x - 8$
$= 3x^2 + 10x - 8$

39. $(2a+4)(3a-5)$
$= 6a^2 - 10a + 12a - 20$
$= 6a^2 + 2a - 20$

41. $(2x+3)^2 = (2x+3)(2x+3)$
$= 4x^2 + 6x + 6x + 9$
$= 4x^2 + 12x + 9$

43. $(2x-3)^2 = (2x-3)(2x-3)$
$= 4x^2 - 6x - 6x + 9$
$= 4x^2 - 12x + 9$

45. $(5t-1)^2 = (5t-1)(5t-1)$
$= 25t^2 - 5t - 5t + 1$
$= 25t^2 - 10t + 1$

47. $(9b-2)^2 = (9b-2)(9b-2)$
$= 81b^2 - 18b - 18b + 4$
$= 81b^2 - 36b + 4$

49. $(2x+1)(3x^2 - 2x + 1)$
$= 2x(3x^2 - 2x + 1) + 1(3x^2 - 2x + 1)$
$= 6x^3 - 4x^2 + 2x + 3x^2 - 2x + 1$
$= 6x^3 - x^2 + 1$

51. $(x-1)(x^2 + x + 1)$
$= x(x^2 + x + 1) - 1(x^2 + x + 1)$
$= x^3 + x^2 + x - x^2 - x - 1$
$= x^3 - 1$

53. $(x+2)(x^2 - 3x + 1)$
$= x(x^2 - 3x + 1) + 2(x^2 - 3x + 1)$
$= x^3 - 3x^2 + x + 2x^2 - 6x + 2$
$= x^3 - x^2 - 5x + 2$

55. $(r^2 - r + 3)(r^2 - 4r - 5)$
$= r^2(r^2 - 4r - 5) - r(r^2 - 4r - 5)$
$\quad + 3(r^2 - 4r - 5)$
$= r^4 - 4r^3 - 5r^2 - r^3 + 4r^2$
$\quad + 5r + 3r^2 - 12r - 15$
$= r^4 - 5r^3 + 2r^2 - 7r - 15$

57.
$$\begin{array}{r} 4x+3 \\ x+2 \\ \hline 8x+6 \\ +\ 4x^2+3x \\ \hline 4x^2+11x+6 \end{array}$$

59.
$$\begin{array}{r} 4x-2 \\ 3x+5 \\ \hline 20x-10 \\ +12x^2-6x \\ \hline 12x^2+14x-10 \end{array}$$

61.
$$\begin{array}{r} x^2-x+1 \\ x+1 \\ \hline x^2-x+1 \\ +x^3-x^2+x \\ \hline x^3+1 \end{array}$$

63.
$$\begin{array}{r} 4x^2+3x-4 \\ 3x+2 \\ \hline 8x^2+\ 6x-8 \\ +\ 12x^3+\ 9x^2-12x \\ \hline 12x^3+17x^2-\ 6x-8 \end{array}$$

APPLICATIONS

65. $A = lw = (x+2)(x-2)$
$= x^2 - 2x + 2x - 4$
$= (x^2 - 4)\, ft^2$

67. $A = lw = (3x-1)(2x+1)$
$= 6x^2 + 3x - 2x - 1$
$= (6x^2 + x - 1)\, cm^2$

69. $A = lw = (7x+3)(5x+4)$
$= 35x^2 + 28x + 15x + 12$
$= (35x^2 + 43x + 12)\, in^2$

WRITING

71. To multiply two binomials, multiply each term of one binomial by both terms of the other binomial, and then combine like terms.

73. They are not equal because, since $(x+1)^2 = (x+1)(x+1)$, the "OI" part of the FOIL process is missed.

REVIEW

75. four and ninety-one thousandths

77.
$$\begin{array}{r} 0.109375 \\ 64\overline{)7.000000} \\ -64 \\ \hline 60 \\ -0 \\ \hline 600 \\ -576 \\ \hline 240 \\ -192 \\ \hline 480 \\ -448 \\ \hline 320 \\ -320 \\ \hline 0 \end{array}$$

79.
$$\begin{array}{r} \overset{1\ 1\ \ \ 1}{56.090} \\ 78.000 \\ +\ \ \ 0.567 \\ \hline 134.657 \end{array}$$

224 Tussy/Gustafson/Koenig Basic Mathematics for College Students, 4e

81. $\sqrt{16}+\sqrt{36}=4+6=10$

Section III.1: Inductive and Deductive Reasoning

VOCABULARY

1. <u>Inductive</u> reasoning draws general conclusions from specific observations.

CONCEPTS

3. circular

5. alternating

7. alternating

9. 10 A.M.

GUIDED PRACTICE

11. add 4 each time: 13 + 4 = 17

13. add 4, add 5, add 6, add 7… : 20 + 7 = 27

15. subtract 3 each time: 6 – 3 = 3

17. subtract 2, subtract 3, subtract 4, subtract 5… : -12 – 5 = -17

19. the pattern: add 4 letters, subtract 1: N + OPQ + **R**

21. the pattern: subtract 1 letter, add 2: f – **e**

23.

25.

27.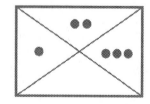

29.

31. The remaining classes are chemistry and zoology. The remaining professors are C and D. Since D doesn't want to teach zoology, D must be teaching chemistry.

33. Since John and Paula are already married, the teacher and the baker must be Luis and Maria. Since Luis is the baker, Maria must be the teacher.

35. The 63 and 47 both include the 16 managers with experience in both. 63 + 47 – 16 = 94 managers with experience in at least one category. There are 100 – 94 = 6 managers with experience in neither category.

37. The 11 and the 15 both include the 8 students with a brother and a sister. There are 11 + 15 – 8 = 18 children with at least one sibling, leaving 27 – 18 = 9 children with no siblings.

TRY IT YOURSELF

39. the pattern: add two letters and change the case: g – h - **I**

41. the pattern: count from the ends of the alphabet, alternating: C – **W**

43.

45. the pattern: subtract a letter, then add 4: L - **K**

47. +16, -15, +14, -13, +12, -11, +10: -4 + 10 = 6

49. -4 , +2 , -4 , +2 … 1 + 2 = 3

51. -1 , -2 , -1 , -2 … -9 – 2 = -11

53. +2 , +1 , -2 , +2 , +1 , -2 … 11 – 2 = 9

55. From point 3: 1,2,3,T

From point 2, since one end is taken: M,2,3,T

Since the lion can't be next to the tiger: M,L,3,T

The zebra is in cage 3.

57. From point 1: B is not last.

From point 2: 1,D,A,C or 1,C,A,D
(since B is not last – note that he must be first!)

From point 3: 1,D,A,C

The placement was B , D , A , C

APPLICATIONS

59. The 35 is included twice in the total.

$997 + 103 - 35 = 1,065$ have served before.

$20,000 - 1,065 = 18,935$ have never served.

61. The 2 is included twice in the total.

$4 + 7 - 2 = 9$ planets are rocky or have moons.

$9 - 9 = 0$ planets have neither.

WRITING

63. Inductive reasoning draws general conclusions from specific observations. Deductive reasoning moves from the general case to the specific.

65. From having watched drivers in the rain, I conclude people don't drive as well when it is raining.